Advances in Analog and Digital Communications

Advances in Analog and Digital Communications

Edited by **Nelson Carter**

CWILLFORD PRESS

New York

Published by Willford Press,
118-35 Queens Blvd., Suite 400,
Forest Hills, NY 11375, USA
www.willfordpress.com

Advances in Analog and Digital Communications
Edited by Nelson Carter

International Standard Book Number: 978-1-68285-138-8 (Hardback)

The publisher's policy is to use permanent paper from mills that operate a sustainable forestry policy. Furthermore, the publisher ensures that the text paper and cover boards used have met acceptable environmental accreditation standards.

Trademark Notice: Registered trademark of products or corporate names are used only for explanation and identification without intent to infringe.

Printed in the United States of America.

Contents

Preface

Analog and digital communications are two distinct yet interrelated means of data transmission. The aim of this book is to present researches that have transformed these fields and aided their advancement. Detailed discussions on various aspects such as protocols, algorithms, integration of networks, performance, modeling, etc. have been covered in this text. It is a vital tool for all researching or studying electronics, telecommunications and similar disciplines. This book gives incredible insights into emerging trends and concepts of this subject. With state-of-the-art inputs by acclaimed experts of this field, this book targets students and professionals alike.

After months of intensive research and writing, this book is the end result of all who devoted their time and efforts in the initiation and progress of this book. It will surely be a source of reference in enhancing the required knowledge of the new developments in the area. During the course of developing this book, certain measures such as accuracy, authenticity and research focused analytical studies were given preference in order to produce a comprehensive book in the area of study.

This book would not have been possible without the efforts of the authors and the publisher. I extend my sincere thanks to them. Secondly, I express my gratitude to my family and well-wishers. And most importantly, I thank my students for constantly expressing their willingness and curiosity in enhancing their knowledge in the field, which encourages me to take up further research projects for the advancement of the area.

Editor

AN ADAPTIVE MULTIMEDIA-ORIENTED HANDOFF SCHEME FOR IEEE 802.11 WLANS

Ahmed Riadh Rebai[1] and Saïd Hanafi[2]

[1]Electrical & Computer Engineering Program, Texas A&M University, Doha, Qatar
riadh.rebai@qatar.tamu.edu
[2]LAMIH Laboratory, University of Valenciennes & Hainaut-Cambrésis, Lille, France
said.hanafi@univ-valenciennes.fr

ABSTRACT

Previous studies have shown that the actual handoff schemes employed in the IEEE 802.11 Wireless LANs (WLANs) do not meet the strict delay constraints placed by many multimedia applications like Voice over IP. Both the active and the passive supported scan modes in the standard handoff procedure have important delay that affects the Quality of Service (QoS) required by the real-time communications over 802.11 networks. In addition, the problem is further compounded by the fact that limited coverage areas of Access Points (APs) occupied in 802.11 infrastructure WLANs create frequent handoffs. We propose a new optimized and fast handoff scheme that decrease both handoff latency and occurrence by performing a seamless prevent scan process and an effective next-AP selection. Through simulations and performance evaluation, we show the effectiveness of the new adaptive handoff that reduces the process latency and adds new context-based parameters. The Results illustrate a QoS delay-respect required by applications and an optimized AP-choice that eliminates handoff events that are not beneficial.

KEYWORDS

IEEE 802.11 WLANs, Inter-cell handoff, QoS constraint, Prevent Scan, next AP-selection, VoIP traffic

1. INTRODUCTION

Recent years have been distinguished by a phenomenal growth in the deployment of the IEEE 802.11 [1] Wireless LANs (WLANs) in various environments like universities [2, 3], companies, shopping centers [4] and hotels. This widespread acceptance can be attributed to decreasing infrastructure costs and potential bandwidth that can be offered to the end user. Many believe that the IEEE 802.11 networks are expected to be part of the integrated fourth generation (4G) networks. However, the limited range of Access Points (APs) causes challenging problems. The Mobile Stations (MSs) are required to find and associate with another AP with acceptable signal quality whenever they go beyond the coverage area of the currently associated AP. The overall process of changing association from one AP to another is called as handoff process and the latency involved in the process is termed as handoff latency.

To meet the lofty goal of becoming the next generation networks, the Quality of Service (QoS) for multimedia applications during handoff should be enhanced. The process must be fast enough to ensure continuous connectivity that may be otherwise prevented by several latency sources incurred at different phases of the handoff process. In 802.11 networks, the handoff process can be divided into three phases: *probing* (scanning), *re-authentication* and *re-association*. According to [5, 7] the handoff procedure in IEEE 802.11 normally takes hundreds of milliseconds, and almost 90% of the handoff delay is due to the search of new APs, the so-called *probe delay*. This rather high handoff latency results in play-out gaps and poor quality of service for time-bounded multimedia applications. On the other hand, The MS association with a specific AP is based only on the Received Signal Strength Indicator (RSSI) measurement of all neighbor APs. The MS will disassociate from the AP when the RSSI falls below a predefined threshold. This procedure is based on the conviction that high RSSI is the best indicator of the quality-of-service provided by the selected AP. This naïve procedure, leads to the undesirable result that many MSs are connected to a few APs, while other neighbor APs remain under utilized or idle. The overloaded APs (with high RSSI) will suffer from performance degradation. This raises the need for a better algorithm that takes into consideration the load on the AP and other context-based parameters, as well as RSSI, as part of MS-AP association.

In this paper, we propose firstly a novel Medium Access Control (MAC) Layer handoff mechanism for IEEE 802.11 networks called Prevent Scan Handoff Procedure (PSHP) that reduces the *probe phase* and adapts the process latency to support most of multimedia applications. The PSHP method decreases the delay incurred during the discovery phase significantly by inserting a new Pre-Scan phase before a poor link quality is observed between the MS and its AP. Based on RSSI measurements, the scanned APs in the pre-scan phase will be sorted in a dynamic list. This new phase will be followed by a "prevent handoff" with a new AP offering better conditions than current AP. As a second proposition, we integrate a new and effective AP-selection layer-2 scheme during the handoff procedure based on Neighbor Graph (NG) manipulation. The proposed technique chooses the next most adapted AP from actual Neighbor APs. This choice is performed by means of a new heuristic function that employs multiple-criteria to derive the search. The new network-configuration method differs from the RSSI constrained process by introducing three new network parameters to optimize the next-AP selection during the 802.11 handoff scan phase. The rest of the paper is organized as follows. Section 2 presents an overview of handoff procedure performed in IEEE 802.11 WLANs and related troubles. Related work found in literature is given and discussed in section 3. In section 4, a detailed explanation of proposed schemes is provided. An experimental analysis of our prototype simulation is shown in section 5 followed by the conclusion in section 6.

2. THE HANDOFF PROCESS IN IEEE 802.11 WLANS

One of the two modes of operation defined in the IEEE 802.11 standard is the infrastructure-based mode. On these widely used infrastructure-based networks, each MS communicates via a special node called AP. If a MS wishes to send or receive data, it first needs to associate with an AP. The AP acts as a bridge and forwards data packets to appropriate destination. Similarly, all the data packets targeted to MSs are passed through their respective APs. Typically, the AP operates on a specific channel and all the MSs need to compete for the channel using one of the access methods described next. Therefore, for an AP with a single transceiver, only one MS will be able to communicate successfully at any specific point of time. The coverage area of an AP is termed as basic service set (BSS). Extended service set (ESS) is an interconnection of BSSs and wired LANs. The logical medium that interconnects BSSs and wired backbone is termed as distributed system (DS) in the literature. It should be noted that wired interface and APs interface to the DS medium are the part of DS. Infrastructure-based WLAN with three BSSs connected with each other by the DS is shown in Figure 1.

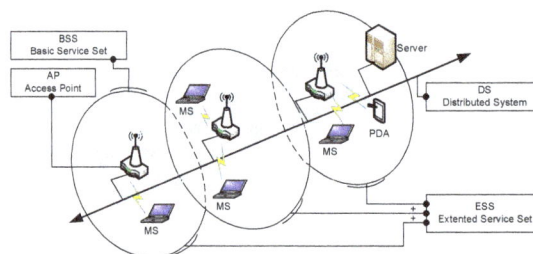

Figure 1. The IEEE 802.11 infrastructure mode

The inter-cell commutation can be divided into three different phases: *detection*, *probing* (scanning) and *effective handoff* (including *authentication* and *re-association*). In order to make a handoff, the MS must first decide when to handoff. However, the IEEE 802.11 standard does not specify any distinct technique to determine when to handoff. The common mechanism is to initiate handoff whenever the Received Signal Strength (RSS) from current AP drops below a pre-specified threshold (termed as handoff threshold in the literature) [5, 8]. Using only current AP's RSS to initiate handoff might force the MS to hold on to the AP with low signal strength while there are better APs in its vicinity. Increasing the handoff threshold does not solve the problem as a larger value drives the MS into performing frequent handoffs. Once the MS decides to make a handoff, the next logical step is to discover the best neighboring AP and re-associate with it. A management frame called *De-authentication* packet is sent, either by the mobile station before changing the actual channel of communication which allows the access point to update its MS-affiliation table, either by the AP which requests the MS to leave the cell. In general, this frame is generated by the mobile station since it detects more quickly the deterioration of the channel quality. After closing the connectivity to the current AP, the MS needs to find potential APs with which to associate. This is accomplished by means of a Medium Access Control layer function called scanning.

There are two types of scanning in the IEEE 802.11 standard: *passive* and *active*. As shown in Figure 2, in the passive scan mode the MS listens to the wireless medium for beacon frames. Beacon frames provide the MS with timing and advertising information. Current APs have a default beacon interval of 100ms [9]. Using information obtained from beacon frames, the MS selects an AP to associate with. During passive scanning, the MS listens to each channel of the physical medium one by one, in an attempt to locate potential APs using the probed channel. Therefore, the passive scan mode incurs significant delay. More technically, the MS commutes from a channel to another one at a regular interval space depending on the setting of *ChannelTime* value. It is indispensable to wait on each channel stated in the *ChannelList* parameter for a time period longer than the *inter-beacon* delays of APs. After scanning all available channels, the MS performs a *Probe* phase (used in active mode) only for the selected AP. As mentioned the polled AP is elected only based on RSSI parameter. The 802.11k group [10] works on improving the choice of the next AP taking into account the network.

In the active scan mode (Figure 3), the MS sends a *Probe Request* packet on each probed channel and waits *MinChannelTime* for a *Probe Response* packet from each reachable AP. If at least one packet is received, the MS extends the sensing interval to *MaxChannelTime* in order to obtain more responses. Contrary, if during *MinChannelTime* the MS does not detect any activity on the channel, the channel is declared *inactive* and the MS passes to the next channel scanning. Thus, the waiting time on each channel is irregular and controlled by two *timers* (not prearranged like the passive scan procedure). When all channels have been scanned, the mobile station collects the information from all available APs and selects the most adequate one to initiate with it the next handoff phase.

The selected AP exchanges IEEE 802.11 authentication messages with the MS. During this phase one of the two authentication methods can be achieved: *Open System Authentication* or *Shared Key Authentication*. The first technique is simply performed by exchanging two authentication packets (request and response) using an access control mechanism based on the MS's MAC physical addresses. It has been shown in [11] the limits of this authentication method by indicating that the access control which is based on mobile's MAC addresses can be easily attacked with software tools that can reconfigure the MAC address of wireless interfaces. The second method assumes the existence of a secret key shared between the station and the access point represented by a *Wired Equivalent Privacy* (WEP) key used also for encrypting data frames. An extra two packets (*challenge - response*) are exchanged during the authentication phase, in which the mobile station must decrypt a text provided by the access point. The method of Shared-Key Authentication requires therefore the exchange of four messages. After that the MS is authenticated by the AP, it sends *Re-association Request* message to the new AP. At this phase, the old and new APs exchange messages defined in Inter Access Point Protocol (IAPP) [12]. Furthermore, once the MS is authenticated, the association process is triggered. The Cell's information is exchanged: the ESSID and supported transmission rates. Only after the association process, the MS will be successfully affiliated with the new AP and can transmit and receive data frames. The initial association starts by exchanging an *Association Request* frame sent by the mobile station that needs to associate. The corresponding AP replays to this request by sending an *Association Response* frame which states whether the association has been accepted or not.

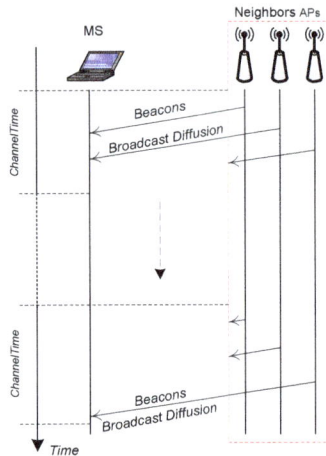

Figure 2. The 802.11 passive scan mode   Figure 3. The 802.11 active scan mode

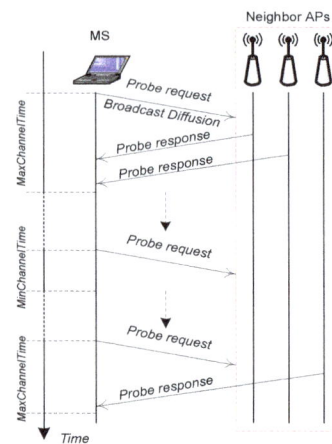

The total delay incurred during these exchanges is referred as the Layer 2 handoff delay, which consists of probe delay, authentication delay and re-association delay. During these various steps, the MS will not able to exchange data with its AP. Based on values defined by the IEEE 802.11 standard, a station will remain inaccessible by any other entity of the network, for nearly 300 to 500 ms [5, 7]. The scanning phase is considered as the most significant contributor to the overall *handoff latency*. An additional process is involved when the MS needs to change its IP connectivity [14]. In such a scenario, the MS needs to find a new access router. Also, the address binding information has to be updated at the home agent and corresponding agent [15]. In our research work we propose an efficient scheme to decrease the latency involved in finding new neighboring AP, which contributes up to 90% of MAC layer handoff delay.

3. RELATED WORK AND RESEARCH OVERVIEW

Numerous schemes have been proposed to reduce the handoff delay in the 802.11 WLANs. In the following, we review the most relevant and representative methods found in the literature. Firstly, we begin with a time investigation for the handoff process to better understand actual difficulties and where to look into. As we have been shown before, the second phase of the handover (*scan* phase) is the most costly in terms of time and traffic. As discussed it is divided into two phases: the *probe* sub-phase and the *channel switching* sub-phase. During these two sub-phases each possible channel must be scrutinized and examined. The latency of the probe sub-phase depends on the adopted scan mode (i.e. passive or active). By assuming the use of a passive scan mode, the average latency of the probe phase depends on the time interval between beacons transmitted periodically by APs and the number of available channels. Explicitly, if the interval between beacons is 100ms of IEEE 802.11b with 11 channels and 802.11a with 32 channels, the average latency will be respectively 1100ms and 3200ms. The switching time incurred while the MS is altering from one channel to another, as it was identified in [16], is negligible and varies between 40 and 150μs. On the other hand, the time incurred with an active scan can be determined by the *MinChannelTime* and *MaxChannelTime* values. Therefore, this quantity can be expressed as shown in Equation 1.

$$N \times MinChannelTime \leq T_{probe} \leq N \times MaxChannelTime \qquad (1)$$

where, N is the number of available channels. The *MinChannelTime* value should be large enough to not miss the *proberesponse* frames and obeys the formula given in Equation 2.

$$MinChannelTime \geq DIFS + (CW \times SlotTime) \qquad (2)$$

where, *DIFS* is the minimum waiting time necessary for a frame to access to the channel. The backoff interval is represented by the contention window (*CW*) multiply by *SlotTime*. In other words, the parameter *MinChannelTime* represents the maximum time for sending a frame. Once this time is elapsed, the MS should receive a response from the access point, and so, will increase the waiting time to *MaxChannelTime* for other potential responses. Otherwise, the station considers that there is no AP on the scanned channel, or other traffic are competing the channel access with the expected management frame.

Regarding the third phase of the handover procedure that allows the MS's identity verification, the delay is changeable. According to the security used, the authentication process can be more or less long. In an untrusted system, only two Authentication frames are exchanged, with their respective 802.11 acknowledgments. Using a secure system, such as WEP, four frames must be exchanged. The latency of the authentication phase is proportional to the number of messages exchanged between the AP and the MS. For example, public systems recently deployed WLAN (e.g. nespot in Korea [17]) use the authentication scheme of the IEEE 802.11x. Therefore, the authentication phase is expected to become an issue much more interesting in future 802.11 versions.

The phase of association or re-association process that comes to the end for the 802.11 handover takes place through the exchange of two frames (Association Request and Association Response), both messages will be acquitted. The duration of this phase, such as the authentication phase, is limited to the medium access time which depends on the traffic in the cell (such management frames have no special priority) and to the delay of their transmissions. In [9] the delay of these last two phases was estimated to less than 4ms in absence of a heavy traffic in the new selected cell. The total handover latency is expressed in Equation 3.

$$T_{Handover} = N \times (T_{switch} + T_{probe}) + T_{authentication} + T_{association} \qquad (3)$$

where, N is the number of available channels depending on the *ChannelList* parameter. In practice, and based on Equation 3, a handover performed on the standard 802.11 network can theoretically have values ranging from 114 ms to 940 ms (for $N = 11$). This value is very high and not acceptable for most of applications with QoS requirement (e.g. voice frames that are time-bounded must be received every 50 ms).

In [18] authors propose an innovative solution to optimize the AP's exploration during the scan phase based on the use of sensors operating on the 802.11 network. These sensors are arranged in cells and spaced 50 to 150 meters.

These sensors have a role to listen to the network using beacons sent periodically by in-range APs. Each sensor is able to identify the nearby access points that are available. When the MS should change its actual cell, it performs a pre-scan operation which involves the sending of a request query to the sensors. Only sensors that have received this request (in range of the MS) react by sending the list of APs that they have identified. Each sensor responds by using a contention window calculated proportionally to the signal strength of the received request message. We figure out that this solution is effective in terms of the next-AP choice and the consequent results have improved significantly the standard handoff scheme. However, it is very expensive and has an extra cost by causing an additional load of unnecessary network traffic due to the sensor use. Moreover, this method is a non compliant solution with the actual 802.11 networks and requires radical changes to adapt it.

In [19], a new handoff scheme called *SyncScan*, is proposed to reduce the probe delay. Unlike the existing probe procedures defined in IEEE 802.11, *SyncScan* allows a MS to monitor the proximity of the nearby APs continuously. In other words, the MS regularly switches to each channel and records the signal strengths of the channels. By doing so, the MS can keep track of information on all neighbor APs. Essentially, this technique replaces the existing large temporal additional costs during the scan phase by a continuous process that passively monitors the presence of access points in other channels. The absence delay of the MS with its current channel is minimized by synchronizing short listening periods of other channels with regular periodic beacon transmissions from each AP. Moreover, through continuous monitoring the signaling quality of multiple APs, a better handoff decision can be made and the authentication / re-association delay can be also reduced.

The authors synchronize the MS with the transmission of beacons from the APs on each channel. By switching regularly and orderly on each channel, the MS reduces its disconnection delay with its actual AP. However, the *SyncScan* process admits a hidden cost. While it removes the scan phase delay, it adds regular additional interruptions between the MS and its actual AP. Specifically, when the MS examines other channels it cannot send or listen to its own AP. As results, the MS may miss packets that were sent when exploring other channels. These errors are very costly in terms of frame loss and performed retransmissions especially for time-bounded applications. Moreover, this extra charge will always affect all MSs even those that will never proceed to a handoff.

In [20], the authors proposed a *selective scan* technique in the IEEE 802.11 WLAN contexts that support the IAPP protocol [12] to decrease the handover latency. This mechanism reduces the scan time of a new AP by combining an enhanced Neighbor Graph (NG) [21] scheme and an enhanced IAPP scheme. If a MS knows exactly its adjacent APs, it can use selective scanning by unicast to avoid scanning all channels. During handoff, a MS does not scan all channels. Instead, it selectively scans few potential APs with unicast based on the NG provided by a NG Server called RADIUS server [22]. They enhanced the NG approach by putting the MS to power-saving mode (PSM) to pre-scan neighboring APs. Then they further derived selective scanning with unicast in power-save mode, pre-registration of IAPP, and frame forwarding-and-buffering mechanisms. Selective scanning allows a MS to only try potential handoff targets. Pre-registration allows early transfer of the MS security context from its old AP to new AP. The forwarding-and-buffering mechanism is to solve the packet loss problem during the handoff process. This solution reduces, in a remarkable way, the total latency of the handoff mechanism. On the other hand, it requires that the MS must have knowledge of the network architecture, it must know exactly the APs which are adjacent for it to be able to employ selective scan and to avoid scanning all channels. In addition, we should take into account the number of packets added by the IAPP that may affect the current traffic. Moreover, we note that all data packets have been sent to the old AP and then routed to the new selected AP before the link-layer is updated, which corresponds to a double transmission of the same data frames in the network. Thus, it greatly increases both the collision and the loss rates in 802.11 wireless networks.

In [23], the authors proposed two changes to the basic algorithm of IEEE 802.11 which reduce significantly the handover average latency using inter-AP communications during the scan phase. In the first proposed scheme, the additional costs incurred during this phase are reduced by forcing the potential APs to send their probe response packets to the old AP and not to the MS which sends the probe request. Therefore, the MS will avoid the waiting delay of *MinChannelTime* or *MaxChannelTime* as performed in the classical IEEE 802.11 approach. Consequently, the MS avoids the packet loss of data without any additional cost in the network. This efficient and fast handoff process is called Fast Handoff by Avoiding Probe wait (FHAP). As explained, the probe wait is avoided by forcing all the neighboring APs operating on the examined channel to send the probe responses to the previous AP using the IAPP protocol. The MS just switches to all the channels and sends the probe request. After the probe phase the MS switches back to its actual AP and it receives the probe responses after sending a request. The discovery phase ends at this time and the MS resumes re-authentication process with the new AP.

Three drawbacks related to the FHAP approach can be noted down. Firstly, the MS should receive packets (probe response messages from potential APs) from its old AP. This implies that the handover threshold must be adjusted so that the MS can communicate with the old AP after the probe phase. Secondly, the problem of non-delivery probe response packets from potential APs to the old AP should be addressed. Finally, as the probe response packets are received via the current AP and not on their respective channels, the MS will not be able to measure the instantaneous values of RSSI and therefore evaluate the quality of the visited channel (which is possible only if the reception is done on the same channel).

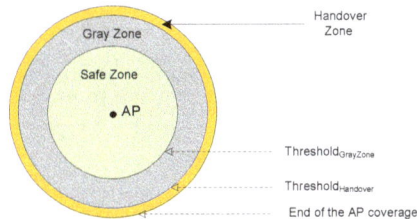

Figure 4. Sub-Zone partitioning in APFH [23]

In [23, 24], the authors have improved their technique FHAP by proposing a new mechanism called Adaptive Preemptive Fast Handoff (APFH). The APFH method requires that the MS predetermines a new AP before the handover begins. Then, the handover threshold is reached, the MS avoids the discovery phase and triggers immediately the re-authentication phase. This process will reduce the total handover latency by decreasing its value to the re-association/authentication delay. Since the authors did not specify how the MS preselects a new AP, we can figure out that the *SyncScan* mechanism [19] presents a solution to this problem. The new adjustments achieved in APFH technique provide a better preemptive scan phase of APs. As shown in Figure 4, the APFH method splits the coverage area of the AP depending on the signal strength in three areas: safe zone, gray zone and handover zone. As its name indicates, the safe zone is the part of the coverage area where the MS is not under a handover threat. Consequently, the MS does not trigger the discovery phase and the data transfer is accomplished normally. The gray area is defined as an area where the handover probability is high. The MS begins collecting information on a new best AP once it enters the unsure zone. The maximum selected speed of MSs for the simulations is 15 m/s as in [25]. In conclusion, since the first proposed scheme FHAP as discussed does not meet the QoS constraints of multimedia applications by receiving all response frames on the old AP, the authors presented a second mechanism called APFH that removes the entire handover latency and respects these strict constraints of VoIP frames.

Many research works were done on the network-layer regarding the challenge to support the mobility in IP networks. New features have been proposed and added to the standard – i .e. IPv6 [14, 15]. However, the best handoff techniques that minimize the scan phase delay are performed at the MAC-layer of the 802.11 standard. By minimizing this phase latency, then the number of lost packets is reduced and the MS will be not reachable only for a limited time.

4. PROPOSED SCHEMES

4.1. Prevent-Scan Handoff Procedure (PSHP)

First, the typical handoff latency in IEEE 802.11b with IAPP network may take a probe delay of 40 to 300ms with a constant IAPP delay of 40ms [26]. To allow the IAPP protocol to reduce this delay, we impose that the MS must authenticate itself with the first AP of the ESS. However, the IEEE 802.11 standard neither requires that authentication must immediately proceed to association nor that authentication must immediately follow a channel scan cycle. The IAPP based pre-authentication [27] is achieved even before MS enters into the discovery state, thus, it does not contribute to the handoff latency.

As a first modification, we propose to define a new threshold other than the existing handoff threshold in the 802.11 standard. We call the new threshold: Preventive Received Signal Strength Indicator which is termed by ($RSSI_{prev}$) and defined in the given Equation 4.

$$RSSI_{prev} = RSSI_{min} + (RSSI_{max} - RSSI_{min})/2 \qquad (4)$$

According to our implementation and the tests that we carried out, $RSSI_{max}$ indicates the best link quality that can exist between the MS and its AP. As its name implies, the $RSSI_{prev}$ is a value of the *link quality* above which the MS is not under the threat of imminent handoff. In the proposed approach, the algorithm starts to detect the mobility of a MS when the RSSI value of the current AP degrades and reaches the $RSSI_{prev}$ threshold, after which the MS starts to seek a new AP which can offer a better link quality.

As described before the best mechanisms – as the *SyncScan* mechanism [19], the proposed selective scan [20], and the APFH technique [23] – are imposing that the MS must predetermine a new AP before the start of handoff. Thus, when the handoff threshold is reached, the MS jumps the discovery phase and starts directly the re-authentication phase. This procedure reduces the overall handoff latency. Therefore, most of the operations related to handoff are executed before that a handoff is triggered, including the selection of the next AP and the transfer of MS's context. For each *SyncScan* procedure, the MS must switch to a specific channel until it receives the corresponding beacon, then it switches back to the original channel. So, for each channel the *SyncScan* latency is given by Equation 5 as follows:

$$SyncScan_{delay} = 2 \times T_{switch} + T_{wait} \qquad (5)$$

where, T_{switch} is the switching delay from one channel to another and T_{wait} is the time required to recover the beacons issued by the APs running on a given channel. The total delay of the handoff scan depends on the number of channels to be scanned.

a)   Association procedure

We present a novel approach that provides an enhanced technique for the *preemptive* scan of APs. Indeed proposed technique requires carrying out a scan (or pre-scan) even before triggering a handoff. We continuously maintain the information concerning at most the best few nearby APs in a dynamic list sorted according to the descending order of the best RSSI values. This list, which is updated after each pre-scan, reduces significantly the scanning delay to nearly zero. Each MS maintains its classified AP list. Using this list, the MS does no longer need to carry out a full scan when a handoff is initiated. Rather, it directly selects the AP in the first position of the AP list and performs an association request. Note that an association request will be accepted only if the RSSI of the first AP, in the dynamic list, has a value greater than both the handoff threshold and the actual RSSI measured with the current AP. In other words, this request is accepted only if the first AP of the dynamic list offers to the MS a better link quality better than offered by its current AP and also sufficient to continue operation without losing connectivity with the other entities of the network. If the association with the first AP fails, then the dynamic list is purged and the MS carries out a new pre-scan cycle.

b) A new Pre-scan process

During a pre-scan process, the MS must switch channels and wait for beacons from potential APs, which produces additional temporal costs composed by switching time between channels and waiting time on each one. Consequently, for each channel we calculate the total time of pre-scan using the following Equation 6.

$$T_{pre-scan} = N \times \left(T_{switch} + T_{wait} \right)$$

(6)

Where, N is the number of available channels, T_{switch} is the switching delay from one channel to another and T_{wait} is the time required to receive the potential beacons on a given channel. Despite T_{switch} and T_{wait} have relatively small values; they are still greater than the maximum retransmission time of 802.11 frames (4ms). Therefore, time-bounded packets may be dropped since the MS is unable to acknowledge them. To overcome this drawback, we modified the MS build-in algorithm to announce entering a Power Saving Mode just before switching channels [28]. This causes the AP to buffer packets until the MS returns to its channel and resets the PSM mode. Since these buffers will not be overfilled during the PSM mode (very short in duration), they are quickly emptied when the MS finishes the pre-can process and returns to normal mode.

The *pre-scan* is programmed so that it does not disturb the existing traffic flow between the MS and its AP. After each execution of the pre-scan, the MS must check its current RSSI value. Once the MS associates with a new AP, then it initiates a pre-scan again. The flowchart in Figure 5 illustrates the new pre-scan procedure of the proposed handoff scheme.

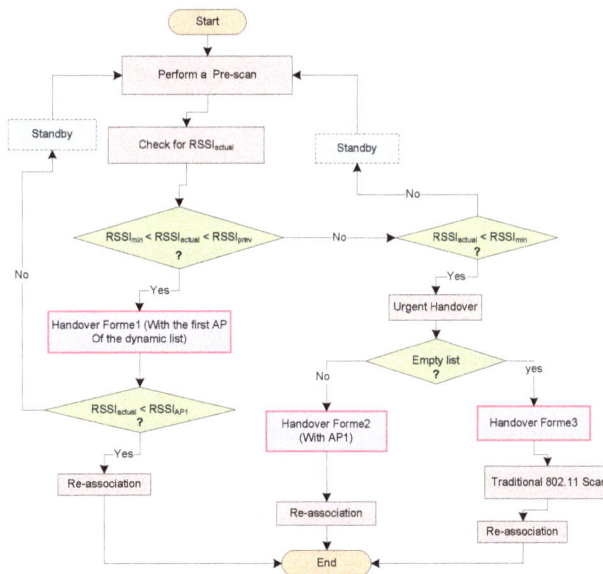

Figure 5. The new PSHP process Flowchart

c)   Operation of the new PSHP mecanism

Figure 6 presents the new state machine for a MS showing the various amendments that we have added to the basic algorithm. Throughout its activity/mobility, the MS can be in one of several states and has various RSSI values. The variation of the RSSI value can be also due to other factors, for example the channel conditions, interference, and AP traffic loads. In the following, we explain our new approach by detailing the various states that a MS may be faced, and conditions that trigger the transition from one state to another. When a MS is initiated in the network, it firstly associates with an active AP. The MS is required to directly proceed to an authentication (called Pre-Authentication) with all other APs in the same ESS. Following the pre-authentication phase and when needed (scan phase), the MS will notify its current AP that it is entering the power-saving mode so that the AP can buffer the incoming data for the corresponding MS. The MS carries out a periodic active scan (each α ms), called a pre-scan phase. We decide that this cyclic pre-scan will not depend on the existing traffic category between the MS and its current AP since it requires a deep cross-layer knowledge of the traffic type. In other words, the MS performs the planned pre-scan mode when lower or higher priority traffic is transmitted on the channel. Otherwise, the proposed mobility technique will have a hidden and costly delay because it can not start until a deep packet classification based on the application data inside IP packets is performed. In fact, carrying out such classification before a pre-scan does not affect the QoS constraints since TCP will retransmit missing packets. As well, the effect may be worse on RTP/UPD traffic. If we choose to trigger a pre-scan mode only when low priority traffic is adopted between the MS and its actual AP and there were QoS packets just transmitted on the network, the pre-scan phase will never happen and our implementation will be not valid and totally worse. Thus, we simplify the algorithm by making the MS enters the PSM mode whenever RSSI crosses the threshold and a planned pre-scan is launched.

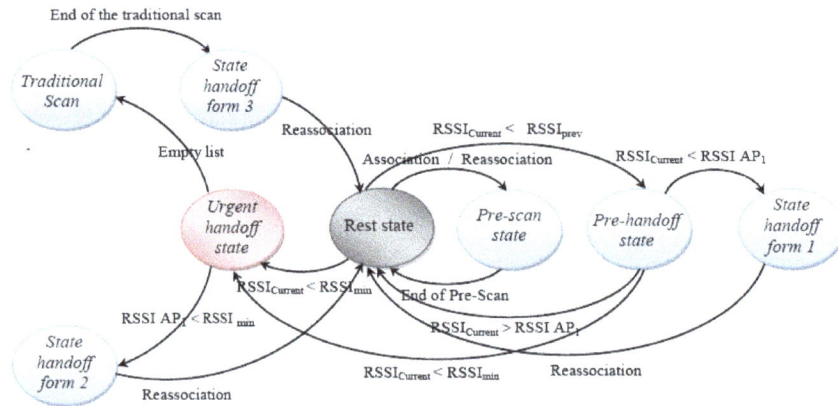

Figure 6. State machine of the PSHP procedure performed by the mobile station

The major advantage of the proposed scheme is that a MS will seek periodically for a new AP offering a better quality of link for forthcoming transmissions between the MS and its associated AP. The periodicity of the pre-scan phase is referred by the parameter α which is defined in the following Equation 7:

$$\alpha = \left[\left(T_{switch} + MaxChannelTime\right) \times N\right] \times 1.5 \qquad (7)$$

Where T_{switch} is the switching delay from one channel to another, *MaxChannelTime* represents the maximum waiting time to collect all potential probe responses from other APs, and N is the number of available channels. The N value varies depending the standard from 11 channels in 802.11g to 32 channels in 802.11a (e.g. N=13 in IEEE802.11b [29]). The periodicity α value is chosen as a manner to ensure that the MS finishes the actual pre-scan cycle and leaves the PSM mode to join back the active mode and receive data packets held by the actual AP before performing the next pre-scan cycle. During the pre-scan period, the proposed algorithm will drive the MS to collect and keep valued information related to each potential AP in the network in a dynamic list. Initially, this list is empty and will be updated after the first pre-scan cycles to include at least one AP with which the MS can be associated if the link quality with its actual AP will attain the handoff threshold. The new technique chooses a maximum of six APs to be saved in the list, since most WLAN infrastructures adopt a hexagonal deployment of AP cells. Thus, the result of the pre-scan cycle is an ordered list of the nearby APs according to their RSSI values. This process is periodically generated to update the dynamic list. Therefore, this repetitive deployment enables the MS to be always reorganized facing to the active network state by keeping a dynamic list linked to the events that occured in the nearest past.

In the proposed handoff scheme we enumerate three forms of handoff that can be happened depending on network conditions. Initially, the MS is in standby state as shown in Figure 6. If the RSSI value associated to the current AP degrades and reaches the $RSSI_{prev}$, then the MS switches to the pre-handoff state to check its dynamic list. It will try to find out a new AP with a corresponding RSSI value higher than the actual one. If such value exists, the MS switches to a 'handoff form1' state and performs a re-association procedure with the chosen AP. Otherwise, the MS returns to its standby state. We notice if such case is achieved ('handoff form1' state) the total latency of the handoff mechanism is reduced to a cost almost equal to that of the re-association delay. If the measured RSSI value with the current AP is deteriorating suddenly and reaches the minimum bound (handover threshold), then the MS passes directly from the standby state to the 'urgent handover' state. In such state the MS must decide whether to perform the second or the third form of handover depending only on the instantaneous data of the dynamic list. If the first AP in the list has a RSSI value greater that the handover threshold, then the MS switches to the 'handoff form2' state. It performs a re-association process with the selected AP and returns to the standby state. If such case does not exist, the MS switches to the 'handoff form3' state in which it carries out a classical 802.11 handoff with a traditional scan procedure. We figure out in 'handoff form2' process, the overall handoff latency is equal also to the re-association delay and the MS chooses an AP from the list which guarantees a minimum channel quality required to transmit data packets. In the third handoff form the MS joins the new AP after executing a standard scan process and returns to the standby state. However, this state is rare and very occasional in practice (the list is rarely empty after carrying out pre-scan cycles).

The main advantage of the proposed technique is its autonomy since it follows instantaneous network variations and takes appropriate decisions accordingly. This allows a faster and more adequate handover occurrence and a channel quality improvement. In addition, the periodic scan presents another opportunity to improve the link quality of the MS with its AP.

4.2. New enhanced technique for a best next-AP selection

In this sub-section, we show a novel and effective layer-2 AP-selection add-on technique for the handoff procedure. The proposed method chooses the most adapted AP from actual Neighbor APs for the next handover occurrence. This choice is performed by means of a new heuristic function that employs multiple-criteria to derive the search. We point out that the standard procedure is based only on the RSSI value considered as the best indicator of the quality-of-service provided by the AP. This naïve procedure, leads to the undesirable result that many MSs are connected to a few APs, while other neighbor APs are underutilized or completely idle. The overloaded APs (with high RSSI values) will suffer from performance degradation. This inefficacity raises the need of a better algorithm which takes into consideration the load on the AP, as well as the RSSI indicator, as part of associating a new MS to the AP. The load balancing problem, as part of handoff, has not been adequately addressed in the literature. In

[30], the authors argued that the login data with the APs can reflect the actual situation of handovers given discrete WLAN deployment. As an example, two WLANs may be very close to each other but separated by a highway or a river. In such case, the user will never move across to the other WLAN. Conversely, if the user is moving fast (e.g. in a train), handover may need to take place among WLANs that are far apart, i.e. among non-neighbor APs. Thus, the user connection history allows us to better predict the probability of the user's next movement.

a) New decisional parameters

We propose a new network-configuration method that differs from the RSSI constrained process [13] by introducing three new network parameters to optimize the next-AP selection during a WLAN handoff procedure. The first parameter is called MS_i and indicates the number of MSs associated with the potential neighbor AP_i. Thus, the new parameter exploits the overload of AP_i as a handoff indicator. A handover occurrence with an overloaded AP may not be beneficial for both the MS and the chosen AP. The second parameter is called CNX_i which is a history-based factor that counts handoff occurrences between the actual AP and the potential next AP_i. This counter is incremented by one each time a handoff occurs between AP and AP_i. This parameter is adopted to select the neighbor AP_i with the maximum CNX_i value as the best candidate for the next handover against other neighbors. It includes the location and other context-based information useful for the next AP-selection. The third parameter is EXT_i which reproduces the number of APs which are neighbors of the potential AP_i chosen for the next handoff process with the current AP. In other words, EXT_i is the number of 2-hop neighbors – denoted by AP_k – of the current AP through a direct neighbor AP_i. This "look-ahead" parameter is added to improve the choice of the potential AP_i to maintain long-term connections and maximize the handoff benefit for new affiliated MSs.

Now, we describe necessary modifications in the MS-AP communication protocol regarding the algorithm implementation. The following new fields must be added to the Beacon and Probe response frames to transmit the additional information needed by the algorithm:

- MS_i, the number of mobile stations (MSs) associated with AP_i.
- CNX_i, the number of actual handoff occurrences between the current AP and desired AP_i neighbor.
- EXT_i, the number of 2-step neighbor AP_k of actual AP through its direct neighbor AP_i.
- $RSSI_i$, the RSSI value of the incoming Probe Request from neighbor AP_i.

As shown in Figure 7, the MS will choose the best AP_i for the actual handoff process after receiving all Probe Responses with the required information.

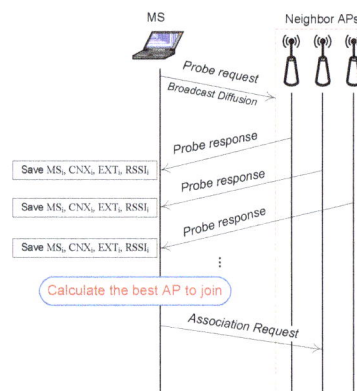

Figure 7. The new AP-selection procedure

b) Mathematical formulation

In this paragraph, we introduce a new numerical optimization approach for the next-AP selection in the IEEE 802.11 handoff procedure. We start by introducing the following assertions:

- A time limit factor is needed because of the MS mobility.
- The next-AP selection can be formed as an *assignment problem*: how to allocate MSs with the given set of APs.
- The measured RSSI values can be different for two MSs allocated to the same AP.

We will adopt the following notation:

- I : is the set of APs
- J : is the set of MSs
- $N(AP_i)$: is the set of the direct neighbors of AP_i
- O_{ik}: is the number of handoffs that are performed between AP_i and AP_k

Then, the matrix A is representing MS-AP affiliations:

$$A_{ij} = \begin{cases} 1 & \text{if the mobile station MS}_j \text{ is associated to the AP}_i \\ 0 & \text{otherwise} \end{cases}$$

We assume that the current mobile station MS_{j° is assigned to the AP_{i°, and our goal is to assign the current MS_{j° to the best new $AP_{i*} \in N(AP_{i^\circ})$ such that:

1) Max $\{RSS_i : i \in N(AP_{i^\circ}) \text{ with } RSS_i \leq Threshold\}$
2) Max $\{|N(AP_i)| : AP_i \in N(AP_{i^\circ})\}$
3) Max $\{O_{i^\circ i} : i \in N(AP_{i^\circ})\}$
4) Min $\{\sum_{j \in J} A_{i,j} : i \in N(AP_{i^\circ}) \text{ with } \sum_{j \in J} A_{i,j} < m\}$ (where m is the maximum load of an AP).

Now we define the variables x_i as:

$$x_i = \begin{cases} 1 & \text{if the current station MS}_{j^\circ} \text{ is assigned to the AP}_i \\ 0 & \text{otherwise} \end{cases}$$

This problem can be formulated as a multi-objective optimization problem:

$$Max \sum_{i \in I} RSS_i x_i$$
$$Max \sum_{i \in I} |N(AP_i)| x_i$$
$$Max \sum_{i \in I} O_{i^\circ, i} x_i$$
$$Min \sum_{i \in I} \sum_{j \in J} A_{i,j} x_i$$

Subject to

$$\sum_{i \in I} x_i = 1$$
$$\sum_{i \in I} RSS_i x_i \leq Threshold$$
$$\sum_{i \in I} \sum_{j \in J} A_{i,j} x_i < m$$
$$x_i = 0 \quad \text{for } i \notin N(AP_{i^\circ})$$
$$x_i \in \{0, 1\} \quad \text{for } i \in I.$$

5. SIMULATION RESULTS AND DISCUSSIONS

5.1. The proposed PSHP evaluation

a) Parameter setting

In this section, the performance of the proposed scheme PSHP is evaluated and compared to the basic handoff scheme (currently used by most network interface cards) and other significant works founded in [9, 19, and 23]. The handoff latencies of all schemes for different traffic loads are presented. This is followed by discussion on the total amount of time spent on handoff for all schemes. The effect of the proposed schemes on real time traffic is explored and weighed against the basic handoff scheme. We used C++ to simulate the new 802.11 handoff versus other described techniques. The IEEE 802.11b [29] networks are considered for testing the schemes. The total number of the probable channels is assumed to be 11 channels (number of all the legitimate channels used in USA for 802.11b). We employed a total of 100 APs and 500 MSs to carry out the simulations. The other parameters are outlined in Table 1.

Table 1. Simulation Parameters

Parameter	Value
Speed of MS	0.1 – 15 m/s
Mobility Model	Random Way Point
MinChannelTime / MaxChannelTime	7/11 ms
Switch Delay	5 ms
Handoff Threshold	-51 dB
Pre-Scan Threshold	-45 dB

In general, all the solutions suggested for optimizing the handoff process aim to reduce the total latency below 50ms [31] mainly for multimedia applications. The proposed PSHP solution aims to be conforming to this restriction by reducing the total handoff delay incurred in 802.11 WLANs. We choose a free propagation model for the mobile stations. Thus, in performed simulations the received signal strength indicator value is based on the distance between a MS and its AP (RSSI-based positioning) as shown in [32]. The relationship between distance separating a MS and its AP and the received signal strength is described in Equation 8.

$$P_r(d) = P_{0\,(transmitting\,power)} - 20\log_{10}\frac{4\pi d}{\lambda}\,[dBm] \qquad (8)$$

Where $\lambda = \dfrac{c}{f}$, f is the used transmission frequency and c is the velocity of light (the propagation speed of waves). The adopted mobility model is based on the model of random mobility "Random Way Point Mobility Model" presented in [33]. The same moving model has been also adopted in other algorithms [19, 20, 21, and 23].

b) Simulation Results

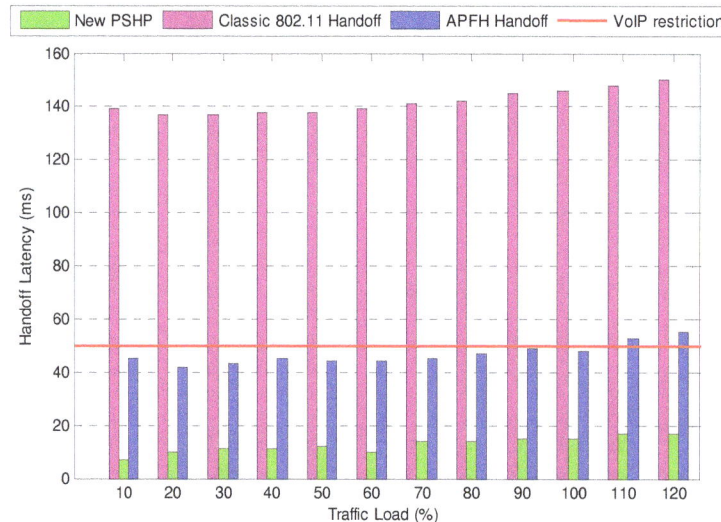

Figure 8. Handoff Latency versus Traffic Loads

Figure 8 shows the average handoff latency against different traffic loads for the three tested schemes. The APFH scheme achieves 67.62% delay improvement while the new PSHP method attains 95.21% improvement versus the basic 802.11 handoff scheme. The handoff latency of the classical approach is consistent with the simulation results in [9] with similar parameters. Also, we point out by observing Figure 8 that the average handoff latencies for the PSHP and APFH schemes are both under 50ms which is well within VoIP constraints. However, PSHP performs the best and the minimal handoff delay compared to the APFH [23]. This remarkable improvement is reached since the new procedure performs a cyclic pre-scan phase before carrying out a handoff and most of handoffs are accomplished early by detecting the premature quality deterioration. As in [19, 23] the traffic load is computed by dividing the number of active MSs (the MSs having data to transmit) over the maximum number of MSs transmitting on one AP's cell. The maximum number of active MSs is equal to 32 in IEEE 802.11 WLANs.

Based on the given results in [5, 19, and 21] of related handoff techniques, we draw the following Table 2 resuming the total handoff delays for corresponding proposed mechanisms. We figure out a significant reduction achieved by the new PSHP algorithm compared to other solutions, and more specifically with the basic handover mechanism. We also note that the solution called *SyncScan* has an important reduction and can also satisfy the time-bounded applications. However, the selective scanning method occasionally exceeds the required QoS limits. This result is due to the inefficacity of the NG graph technique to manage all network topology changes due to the continuous MS mobility. Regarding the new handoff method, and as expected, the total latency is reduced only to the re-authentication phase ($\approx$11ms). This delay can reach more (18ms) because in some simulated cases a handoff occurrence is triggered while a pre-scan cycle did not finish.

Table 2. Average latencies of different handoff procedures

Scan Technique	Total Latency
SyncScan [19]	40±5ms
Selective Scan [21]	48±5ms
APFH [23]	42±7ms
Traditional 802.11 handoff	From 112ms up to 366ms
Proposed PSHP	11±7ms

Figures 9 and 10 evaluate the performance of the APFH technique – the best known solution in literature – and the new PSHP scheme against VoIP traffic. The packet inter-arrival time for VoIP applications is normally equal to 20ms [34], while it is also recommended that the inter-frame delay to be less than 50 ms [31, 34]. This restriction is depicted as a horizontal red line at 50ms in Figures 9 and 10. A node with VoIP inter-arrival time is taken and the corresponding delays are shown. The vertical green dotted lines represent a handoff occurrence. The traffic load for the given simulations was fixed to 50% and the number of packets sent to 600 ($\approx$ 2.5 s).

We remark that handoff occurrences are not simultaneous for the two simulated patterns. The MSs adopting the new PSHP algorithm detect the quality deterioration with their corresponding AP earlier than the APFH process. We note that both techniques respect the time constraint of real-time applications on recorded inter-frame delays without exceeding the required interval (50 ms). However, this constraint is better managed by the new approach and the inter-packet periods are more regular and smaller. As discussed before, the handoff latency for PSHP scheme is just the re-authentication delay if all handovers occur under the first or second form. If the third form of handover is performed, then the latency will be equal to the delay incurred in legacy 802.11 scanning all channels in addition to the re-authentication delay. However, most of PSHP handoff occurrences are carried out using the first and the second form.

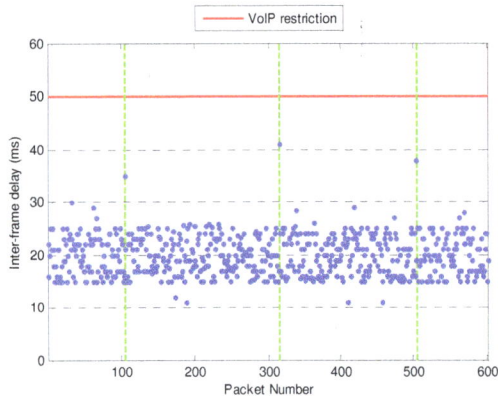

Figure 9. Inter-frame Delay in APFH [23]

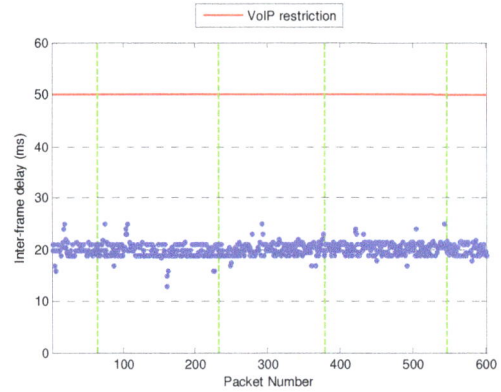

Figure 10. Inter-frame Delay in PSHP

To ensure this last assertion we present in Figures 11 and 12, respectively, a count of handoff occurrences for both APFH and PSHP schemes according to the traffic load and the detailed number of the various handoff forms related to the new PSHP technique. We set the simulation time to 10s for each considered traffic load.

By comparing values obtained by the two algorithms in Figure 11, we easily point out that the APFH technique [23] performs less handovers in the network than the proposed PSHP scheme. This result can be explained by the adoption of the new form of preventive Handover (called Form 1). Using this new form, a MS will not wait for a minimum quality recorded equal to the handoff threshold to trigger a handover. This new technique detects early the link quality deterioration with its current AP and performs a switching with a new AP which improves link conditions. Therefore, the periodic pre-scan adopted by the new technique offers new opportunities to enhance the link quality between a MS and its AP and a significant reduction of the total handover delay. Indeed, with the pre-scan cycle the MS can discover other APs that have a better value of RSSI than provided by the current AP and provide the means to make more intelligent choices before and during a handover. The new algorithm PSHP has a better choice for the next AP by collecting periodic RSSI measurement. Thus, the decision is earlier and more beneficial when a handover is performed (rather than relying on a single sample as in usual schemes). Consequently, the extra number of PSHP occurrences versus APFH procedure happenings is compensated by an early choice of next AP with a better offered quality.

In Figure 12, the vertical red lines represent the executed number of Form 1 handoffs. Blue lines represent the number of handoffs taken under the second and third form, i.e. urgent handoffs. Recall that handoff under the first form is started when the RSSI value degrades below the $RSSI_{prev}$ and above the handoff threshold. Handoffs of the second and third form start only if RSSI value is degraded below the handoff threshold. In Figure 12 we figure out for most traffic loads, urgent handoffs occur less frequently than handoffs of Form 1. We also state that the proposed algorithm presents true opportunity to improve link quality since most of handoff occurrences are executed before that the RSSI value degrades below the handoff threshold. Accordingly, we conclude that almost half of accomplished handoffs are done under the new first form, which explains the delay reduction of PSHP since the first form decreases the related latency considerably and improves the link quality between the MS and its current AP.

Table 3 shows the average probability of data packets being dropped and caused mainly by handoff procedure for the three schemes (APFH, PSHP, and the classic 802.11 approach). We also add the obtained result in [19, 21] for *SyncScan* and *SelectiveScan*, respectively. For comparison purposes, the traffic load for all nodes is divided into real-time and non real-time traffic with a ratio of 7.5/2.5. Other than errors caused by handoff occurrences, the real-time data packets are dropped also if the inter-frame delay exceeds 50ms. The simulation time for each traffic type is 10s (equivalent to about 2500 frames). Clearly, PSHP outperforms the other three schemes and the basic 802.11 as long as the traffic load is limited. The loss probability value of the new PSHP technique is divided by two compared to these obtained by *SyncScan* and *SelectiveScan* methods and by three of that accomplished by the standard 802.11 scheme.

In conclusion, periodic scanning also provides the means to make more intelligent choices when to initiate handoff. The new implementation can discover the presence of APs with stronger RSSIs even before the associated AP's signal has degraded below the threshold. In addition, the pre-scan phase does not affect the existing wireless traffic since the corresponding MS will carry out a pre-scan cycle after declaring the PSM mode to buffer related packets.

5.2. Evaluation of the new add-on AP-selection heuristic

As mentioned above we add new context-based parameters for the next AP choice when a handover is triggered in the network by a MS. The result technique is not dependent on the used handoff method. Thus, we integrate the new developed heuristic function with both the classic and the proposed switching algorithm. Specifically, in the standard 802.11 method the next AP selection will be performed after the scan phase on the found APs by choosing one based on the new objective function. Regarding the PSHP procedure this choice will be performed after each pre-scan cycle only on APs that belong the associated dynamic list. This function is also performed for both handoff Form 2 and Form 3. The only algorithm modification in PSHP Form 1 handoff process is that the objective function is performed only on listed APs that have an RRSI value greater than the actual RSSI measured between the MS and its actual AP. By adopting this condition we always maintain the main purpose of the PSHP which is an earlier selection of a new AP that offers a better link quality. Therefore, the modified PSHP will not choose automatically the first best AP in the list. However, it will select from existing AP that maximizes the objective function and also offers a better channel link quality. We set the same simulation parameters as given in Table 1. However, we add geographic constrains by influencing some MS-AP link qualities depending on AP initial positions and by introducing initial specific values for the CNX parameter to illustrate the already performed MS-journeys in the network and a random primary associations between MSs and the given set of APs. The simulated mobility model regarding the MS moves is no longer "Random Way Point". To be closer to realistic networks and to better assess our mechanism we switch to the "Random Direction" Mobility Model which forces mobile stations to travel to the edge of the simulation area before changing direction and speed. We choose this model because of its inclusion simplicity and instead of the "City Section" Mobility Model – which represents streets within a city. By including these constrain, we evaluated of the proposed heuristic combined with handoff schemes. In Figure 13 we resume the handoff occurrences for both classic and modified handoff schemes for the standard 802.11 and the PSHP techniques according to the traffic load. We set the simulation time to 10s for each considered traffic load. We point out a perceived reduction for handoff occurrences for both schemes when using the proposed heuristic procedure during the next AP selection. The produced results with the PSHP procedure are clearly enhanced in term of handoff count by integrating the new add-on heuristic technique. This is the use effect of the new objective function that accomplishes a better AP choice for the next inter-cell commutation, and consequently, improves the total number of handoff happening by reducing worse AP selections that was based only on RSSI-measurement decisions.

The detailed number of the various handoff forms related to the extended PSHP technique is shown in Figure 14. As well as in Figure 12, the vertical red and blue lines represent, respectively, the executed number of Form 1 handoffs and the count of handoffs taken under the second and third form (called also urgent handoffs). We figure out that handoffs Form 1 – performed when the RSSI value degrades below the $RSSI_{prev}$ threshold – are more triggered using the modified PSHP. We note that the proposed algorithm detects earlier the MS path and direction based on supplementary context-based information, and as a result, chooses quicker the best AP that improves the link quality and offers a continuous channel connection. Accordingly, 72% of accomplished handoffs are done under the first form of PSHP that decrease considerably the total latency and improves the link quality. As discussed before, data packets are dropped mainly by the handoff procedure and the violation of VoIP restrictions. Table 4 shows the data loss average probability for both classic 802.11 and PSHP approaches. As settled before the simulation time is 10s. The traffic load for MSs is equally combining real-time and non real-time traffic. The given results are the average of simulated values by varying the traffic load (from lower to higher loads).

Table 3. VoIP packet's loss Table 4.

Scan Technique	Loss Probability
SyncScan [19]	0.92 x1E-02
Selective Scan [21]	1.28 x1E-02
APFH [23]	0.72 x1E-02
IEEE 802.11 handoff	1.62 x1E-02
New PSHP	0.53 x1E-02

Table 4. Packet's loss with heuristic selection

Scan Technique	Loss Probability
Standard 802.11 handoff	1.62 x1E-02
PSHP	0.53 x1E-02
IEEE802.11+heuristic selection	0.78 x1E-02
PSHP + heuristic selection	0.32 x1E-02

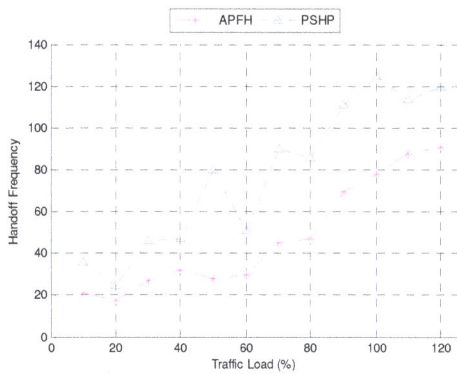

Figure 11. Handoff Frequency

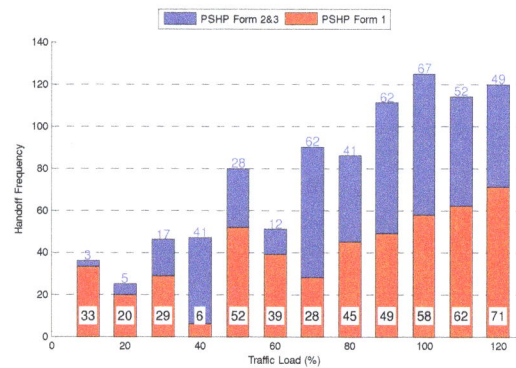

Figure 12. Occurrence of Handoff forms in PSHP

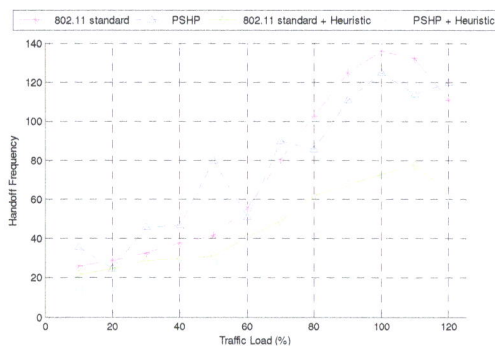

Figure 13. WLAN Handoff's frequency

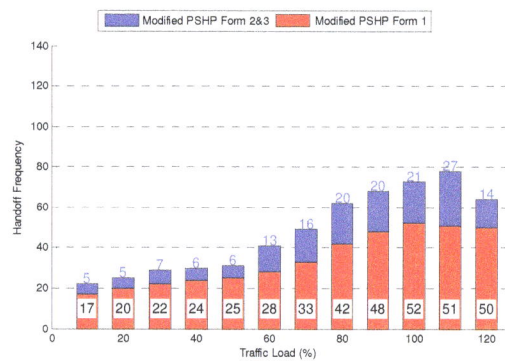

Figure 14. Handoff Occurrence in PSHP

We note that the modified PSHP version is outperforming the regular scheme. The reduced number of handoffs and also the high percentage of Form 1 handoffs lead to minimize the packet loss caused by handoff procedures. Thus, we can conclude that the loss probability value obtained by the new PSHP integrating the heuristic technique includes mainly dropped packets associated to a higher traffic load and not linked to the lack of respect of QoS constrains.

6. CONCLUSIONS

Mobile voice applications are currently the challenge for 802.11-based WLANs. One of the major impediments is the high cost of handoff as MSs room between APs in an infrastructure network. In this paper, we firstly presented a new technique, called PSHP, which reduces the delay and the traffic generated by the handoff process. As demonstrated, the continuous scanning PSHP technique offers significant advantages over other schemes by minimizing the time during which an MS remains out of contact with its AP and allowing handoffs to be made earlier and with more confidence. The result is a staggering 95% reduction of handoff latency compared to the typical procedure. As a second contribution we took into account additional network-based parameters to drive a better next-AP choice. This new add-on profit function is used to insert new factors reflecting resource availability, location, and other context-based information. Thus, the overall network performance is improved by electing from available APs, the one that increases the benefit of the next handoff occurrence.

REFERENCES

[1]   IEEE Std. 802.11-1999, Part 11: Wireless LAN Medium Access Control (MAC) and Physical Layer (PHY) Specifications, IEEE Standard 802.11, 1999.

[2]   D. Corner, J. Lin, and V. Russo, "An Architecture for a Campus-Scale Wireless Mobile Internet," Tech. Rep. CSD-TR 95-058, Purdue University, Computer Science Department.

[3]   A. Hills and D. Johnson, "A Wireless Data Network Infrastructure at Carnegie Mellon University," IEEE Personal Communications, vol. 3, pp. 56–63, Feb. 1996.

[4]   P. Bahl, A. Balachandran, and S. Venkatachary, "Secure Wireless Internet Access in Public Places," in Proceedings of IEEE International Conference on Communications 2001, vol.10, pp. 3271-3275, June 2001.

[5]   A. Mishra, Min ho Shin, W. Arbaugh, "An empirical analysis of the IEEE 802.11 MAC layer handoff process," ACM SIGCOMM Computer Communication Review, vol. 33: 93-102, April 2003.

[6]   Skype software, available at http://www.skype.com/

[7]   G. Bianchi, L. Fratta, and M. Oliveri, "Performance Evaluation and Enhancement of the CSMA/CA MAC Protocol for 802.11 Wireless LANS," Proceedings of the 7th International Symposium on Personal, Indoor and Mobile Radio Communications (PIMRC), vol. 2, pp. 392 – 396, 1996.

[8]   M. Raghavan, A. Mukherjee, H. Liu, Q.A. Zeng, and D. P. Agrawal, "Improvement in QoS for Multimedia Traffic in Wireless LANs during Handoff," Proceedings of the 2005 International Conference on Wireless Networks (ICWN'05), Las Vegas, USA, pp. 251-257, Jun 27-30, 2005.

[9]   H. Velayos and G. Karlsson,"Techniques to Reduce IEEE 802.11b MAC Layer Handover Time," in Proc. IEEE ICC 2004, vol. 7, pp. 3844-3848, June 2004.

[10]   IEEE Standard 802.11k, "Radio Ressource Management", IEEE Standard 802.11- 2003.

[11]   A. Mishra, M. Shin, N. Petroni, T. Clancy, and W. Arbaugh, "Proactive Key Distribution Using Neighbor Graphs," IEEE Wireless Communications Magazine, vol. 11, pp. 26-36, Feb. 2004.

[12]   "IEEE Trial-use Recommended Practice for Multi-Vendor Access Point Interoperability via An Inter-access Point Protocol across Distribution Systems supporting IEEE 802.11 Operation," IEEE Std 802.11F, July 2003.

[13]   M. Shin, A. Mishra, and W. A. Arbaugh, "Improving the Latency of 802.11 Hand-offs using Neighbor Graphs," Proceedings of the ACM MobiSys Conference, Boston, MA, USA, pp. 70-83, June 2004.

[14]   D.Johnson, C.Perkins, and J.Arkko, "Mobility Support in IPv6". RFC 3775 (Proposed Standard): ftp.rfc-editor.org in-notes rfc3775.txt, June 2004.

[15]   T. Cornall, B. Pentland, and P. Khee, "Improved Handover Performance in Wireless Mobile IPv6," Proceeding of the 8th International Conference on Communication Systems (ICCS), vol. 2, pp. 857-861, Nov. 2002.

[16]   IEEE 802.11i, Part 11: Wireless LAN Medium Access Control (MAC) and Physical Layer (PHY) Specifications: Medium Access Control (MAC) Security Enhancements. Supplement to IEEE 802.11 Standard, June 2004.

[17]   Y., S. Park, S. Choi, G. Lee, J. Lee, and H. Jung, "Enhancement of a WLAN-Based Internet Service", ACM Mobile Networks and Applications, vol. 10, no. 3, June 2005, pp. 303-314.

[18]   S. Waharte, K. Ritzenthaler and R. Boutaba, "Selective active scanning for fast handoff in wlan using sensor networks", IEE International Conference on Mobile and Wireless Communication Networks (MWCN'04), Paris, France, October 2004.

[19]   I. Ramani and S. Savage, "SyncScan: Practical Fast Handoff for 802.11 Infrastructure Networks," Proceedings of the IEEE Infocom, vol. 1, pp. 675-684, March 2005.

[20]   P.J. Huang, Y.C. Tseng, K.C. Tsai, "A Fast Handoff Mechanism for IEEE 802.11 and IAPP Networks," The 63rd IEEE Vehicular Technology Conference, 2006. VTC Spring - 2006.

[21]   H. Kim, S. Park, C. Park, J. Kim, and S. Ko, "Selective Channel Scanning for Fast Handoff in Wireless LAN using Neighboor Graph," ITC-CSCC 2004, July 2004.

[22]   Radius, RFC 2865 et 2866, http www.ietf.org rfc rfc2865.txt, http www.ietf.org rfc rfc2866.txt

[23]   V. M. Chintala and Q.A. Zeng, "Novel MAC Layer Handoff Schemes for IEEE 802.11 Wireless LANs". The IEEE Wireless Communications and Networking Conference 2007, WCNC 2007, Mars 2007.

[24]   P. Roshan and J. Leary, "802.11 Wireless LAN Fundamentals, CISCO Press", ISBN No.1587050773.

[25]   M.Raghavan, A. Mukherjee, H. Liu, Q-A. Zeng, and D. P. Agarwal, "Improvement in QoS for Multimedia Traffic in Wireless LANs during Handoff," Proceedings of the 2005 International Conference on Wireless Networks ICWN'05, Las Vegas, USA, pp. 251-257, June 27-30, 2005.

[26]   B. Aboba, "Fast handoff issues," IEEE-03-155rO-I, IEEE 802.11 Working Group, Mars 2003.

[27]   Orinoco TB 034/A, "Inter Access Point Protocol (IAPP)," Technical Bulletin, Feb. 2000.

[28]   S. Baek, and B.D. Choi, "Performance analysis of power save mode in IEEE 802.11 infrastructure WLAN," International Conference on Telecommunications 2008, St. Petersburg, Russia, ICT 2008.

[29]   IEEE Std. 802.11b, Supplement to Part 11: Wireless LAN Medium Access Control (MAC) and Physical Layer (PHY) specifications: Higher-speed Physical Layer Extension in the 2.4 GHz Band, IEEE Std. 802.11b-1999, 1999.

[30]   C.T. Chou and K.G. Shin, "An Enhanced Inter-Access Point Protocol for Uniform Intra and Intersubnet Handoffs", IEEE Transactions on Mobile Computer, vol. 4, no. 4, July/August 2005.

[31]   International Telecommunication Union, "General Characteristics of International Telephone Connections and International Telephone Circuits," ITU-TG.114, 1988.

[32]   T. Kitasuka, K. Hisazumi, T. Nakanishi and A. Fukuda, "Positioning Technique of Wireless LAN Terminals Using RSSI between Terminals", Proc. the 2005 International Conference on Pervasive Systems and Computing (PSC-05), pp. 47-53, Las Vegas, Nevada, USA, June 2005.

[33]   J.Y. Boudec, "On the Stationary Distribution of Speed and Location of Random Waypoint", the IEEE Transactions on Mobile Computing, vol. 4, pp. 404-405, Jul/Aug, 2005.

[34]   Y. Chen, N. Smavatkul, and S. Emeott, "Power management for VoIP over IEEE 802.11 WLAN," Proceedings of the IEEE WCNC 2004, vol.5, pp.1648–1653, March 2004.

2

Cloud-Based Mobile Video Streaming Techniques

Saurabh Goel

Software Engineer, Pariksha Labs Pvt. Ltd. India
Saurabh.goyal6@gmail.com

ABSTRACT

Reasoning processing is changing the landscape of the electronic digital multi-media market by moving the end customers concentrate from possession of video to buying entry to them in the form of on-demand delivery solutions. At the identical time, the cloud is used to collect possessed video pathway and form way out that assist viewers to find a whole new variety of multi-media. Cellular devices are a key car owner of this change, due to their natural mobility and exclusively high transmission rate among end customers. This document investigates cloud centered video streaming methods particularly from the mobile viewpoint. The qualitative part of the research contains explanations of current video development methods, streaming methods and third celebration cloud centered streaming solutions for different mobile which shows my realistic work relevant to streaming methods with RTMP protocols family and solutions for iPhone, Android, Smart mobile phones, Window and BalackBerry phones etc.

KEYWORDS

QCIF, CIF, 4CIF, HD, FFMPEG Encoding/ Streaming, Zencoder cloud based Encoding API , Amazon Cloud Front service, Video Streaming, H.264, MPEG- 4, RTMP, RTMPT, RTMPE, RTMPTE.

1. INTRODUCTION

Developing multi-media content for effective indication over reasoning of cloud based centered mobile system with limited data rates, such as the 3G-324M system needs skills and knowledge. It needs an knowing of the fundamentals that have an effect on movie quality, such as codec choice and compression, and the use of specific resources, such as the FFMPEG Development, and Zencoder Cloud centered Development API which can be used to validate that the material of videos clip data file are effectively specified for end customers.

2. VIDEO FUNDAMENTALS

Due to bandwidths of mobile networks are limited, video data must be encoded/compressed considerably. This part wraps the fundamentals of encoded video and its characteristics within different networks.

2.1. Bandwidth

In multi-media streaming programs, video encoding is used for the reason that uncompressed video needs huge information space to store data. In fact, High definition (HD) films on DVD or Blu-ray are already in a compacted format that provides information of 4 - 6 Megabyte per second. For cellular streaming systems, which can require information rates as low as 30 kilobytes per second, this means that it clip must be compacted thousands of times or more to achieve the required information. With the growth of cellular multi-media streaming, you should work within the information restrictions of the network and the ability of the endpoint. Although 3G and next generation systems provide much higher bandwidths to cellular phone devices, as more and more endpoints use these systems for multi-media projects, conformance to focus on end customers bitrates will become more essential than today [1].

2.2. Networks for Video Streaming

Table 1 illustrates the network atmospheres used for distributing video services with different aspects.

Table 1. Networks Used for streaming Video Services

Network	Bandwidth	Terminals	Codecs	Image Size
3G-324M	64 Kbps	Video Handsets	H.263,MPEG-4,H.264	QCIF,CIF
3G Wireless	256-768 Kbps	Video handsets, smart phones	H.263, H.264, MPEG-4	QCIF, CIF
Broadband IP	768 Kbps	Smart phones, soft client on PC	H.264	QCIF, CIF
Enterprise	2-5 Mbps	Soft client	H.264	CIF, 4CIF, HD
WiMax, LTE	2-100 Mbps	PC, TV, portable devices	H.264	CIF, 4CIF, HD

2.3. Audio/Video Codecs

An audio codec is a system applying criteria that encode and decode electronic digital audio information according to a given sound extendable or movies online sound structure. The item of the criteria is to signify the great stability sound indication with lowest bitrates while protecting the excellent.
Examples: AAC, ADPCM, MP3, WMA, PCM, Vorbis, Dolby AC-3.

A video codec, brief for Encoder/Decoder, is used to encode video information to accomplish a very low bitrate.
Examples: MPEG-2, H.263, MPEG-4 and H.264.

To accomplish such small bitrate audio/video, codecs make use of both lossless and lossy compression methods. We can accomplish this by third celebration system like FFMPEG open source libraries and Zencoder cloud-based encoding API.
Compression performance is the capability of a codec to encode or decode more video/audio features into an information flow described by fewer bits. The better the quality and sharpness of the multimedia clip by using useful codec compression.

3. VIDEO STREAMING

In streaming procedure, it clip data file is sent to the end individual in a (more or less) continuous flow. It is simply a strategy for shifting information such that it can be prepared as a stable and ongoing flow and it is known as Streaming or encoded movie that is sent across information system is known as Streaming. Streaming movie is a series of "moving images" that are sent in compacted form over the Internet and shown by the audience as they appear [4]. A end user never hang on to obtain a large data file before viewing it clip or enjoying the sound.

3.1. Streaming Principle

Real-time video applications require media packets to arrive in a timely manner; excessively delayed packets are useless and are treated as lost [6]. In streaming programs it is necessary for the information packets to reach their location in regular basis because the wait can cause the network blockage, and can result in the decrease in all those packets suffering from extreme wait. This causes decrease in quality of information, the synchronization between customer and hosting server to be damaged and mistakes to distribute in the provided movie.

Two kinds of steaming are, real-time and pre-recorded streaming. User Datagram Protocol (UDP) is used for streaming which delivers the multi-media flow as a sequence of small packets [4]. The majority of transport protocols perform over an RTP stack, which is implemented on top of UDP/IP to provide an end-to-end network transport for video streaming [2].

3.2. Video Streaming Architecture

A cloud based mobile movie streaming scheme is represented in Figure 1. [3]. A cloud based source implements a streaming hosting server which is responsible for retrieving, sending and adapting it clip flow. Depending on the application, it clip may be protected on-line for a real-time broadcasting or pre -encoded and stored for broadcasting an on demand [3].

Programs such as interactive movie, live broadcast, mobile movie streaming or interactive online games require real -time encoding. However, applications such as movie on-demand require pre-encoded movie. When the multicast session is initialized, the streaming hosting server retrieves the compressed movie and begins the loading with the adequate bitrate stream.

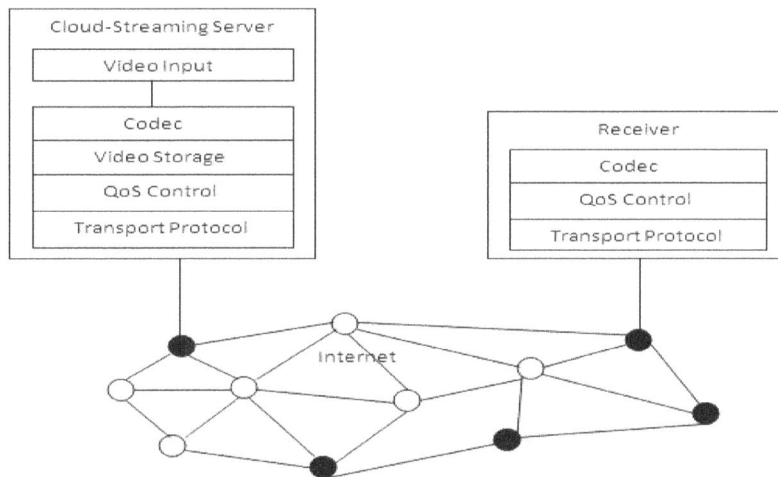

Figure 1. Video Streaming Architecture

4. VIDEO ENCODING TECHNIQUES

Video codecs employ a range of encoded/decoded methods to fit videos signal into the allocated channel bandwidth. These encoding methods can influence the generating quality of it differently. An understanding of development concepts can help a material provider determine what material will look best on a mobile phone, and emphasize some of the expected tradeoffs when generating multi-media data files.

Rapid bandwidth decrease can be carried out by using video encoded/decoded methods such as [1]:

    a. Eliminating mathematical redundancies
    b. dropping quality size (CIF to QCIF)
    c. Using less frames per second (15 fps to 10 fps)

Further bandwidth decrease can be carried out by utilizing the styles within it information and eliminating redundancies. Image compression depends on removing information that is indiscernible to the audience. Motion settlement provides interpolation between frames, using less information to signify the change. The objective of videos encoder/decoder is to take out redundancies in it flow and to scribe as little information as possible. To achieve this objective, the encoder examples it flow in two ways:

    a. In time durations from successive frames (temporal domain)
    b. Between nearby pixels in the same frame (spatial domain)

A video decoder pieces it flow together by treating the development process. The decoder reconstructs it flow by adding together the pixel variations and shape variations to form complete video. In current video encoding principles requirements such as MPEG and H263 families.

4.1. Encoded Video Stream

An encoded video stream consists of two types of encoded frames [1]:

4.1.1. I-Frames

An I-frame is encoded as a single image, without referencing to any other frames. Each 8x8 block is first transformed from the spatial domain into the frequency domain [5]. This is known as a key frame, for the reason that it signifies the referrals key of it video content flow. All pixels that describe the image are defined in the I-frame. Videos clip decoder must begin with an I-frame to decode it clip flow because without an I-frame, a movie decoder has no referrals to determine how movie pixels have changed as the earlier frame. For this reason, compressed movie recordings normally do not begin until an I-frame is received by the videos device.

4.1.2. P-Frames

A P-frames is encoded relative to past reference frame [5], which can either be an I-frame or a before P-frame. The quantity of information in a P-frame is many times small than the quantity of information in an I-frame. If videos clip begins understanding on a P-frame at an endpoint, an individual might see either scrambled movie or no movie, because there is no referrals frame.

4.2. Video Streaming package (.MP4, .3GP)

When streaming multi-media files to cellular handsets, it clips and audio data must be placed in the proper structure. The package structure for cellular multi-media streaming is the .3gp, defined by the 3rd Generation Partnership Project (3GPP) [1] and .mp4 file for delivery to cellular phone devices. For the reason that the bandwidths of multimedia telephone systems networks are confined, Multimedia data included in a .3gp file is compressed considerably. Within the .3gp package, movie can be encoded with specific movie codecs specified by the 3GPP. FFMPEG Encoding and Zencoder cloud based Encoding API support .3gp, .mp4 files with the H.263, MPEG-4, and H.264 movie codecs.

Table 2. Presents an overview of different versions of two standard families

Standards	Applications	Bit rate
H.261	Video teleconferencing over ISDN	64 Kbs
MPEG- 1	Video on digital storage media (CD-ROM)	1.5 Mbs
MPEG- 2	Digital TV	2-20 Mbs
H.263	Video telephony over PSTN	>34 Kbs
MPEG- 4	Multimedia over internet, Object based coding	Variable
H.264/MPEG- 4	Improved video compression	10's-100's Kbs

4.3. Video Streaming limitations

Multimedia streaming is confined by the network channel potential, 3G-324M channel bandwidth, Multi-coded stream, Transcoding, Packet loss, Bandwidth supervision and endpoint features.

5. VIDEO STREAMING TECHNIQUES

There are various streaming techniques for different mobiles, Smartphone describe below:

5.1. Progressive Download

The mobile customers have the choice to gradually get a compressed data clip partitioned in the appropriate codecs for the product to play by using HTTP or HTTPS. As the data file starts to gradually download, play-back is started enabling an almost immediate watching of the material [8]. In the qualifications, the press gamer is constantly on the download the rest of the material. By comparison, without modern download the user would have to wait for the whole data file to obtain to the product before watching would start. During the play-back process, audiences are able to seek back and forth through the whole press data file. If the audience looks for forward to a point in the schedule that has not yet downloadable, the press gamer stop play-back until the data comes.

5.2. HTTP Live Streaming

Hyper text transport protocol (HTTP) structured multimedia streaming communications protocol carried out by Apple company is known as Hyper text transport protocol (HTTP) Live Streaming (HLS).For Apple company products like IOS, Ipad and Iphone etc.,this is an adaptive streaming multimedia distribution standard protocol. It is an exemplified and segmented in MPEG family transport channels and M3U8 - MP3 Playlist File (UTF-8) to offer live and on-demand multimedia data by utilizing H.264 multimedia codec. On the behalf of most suitable channel or stream like bandwidth, platform and CPU limits selected by device instantly, it downloads available bits for buffering to play multimedia file.

HLS streaming provides the best user experience, but its benefits also include good IT practices and important business considerations:

1) The best user experience - There are different formats of multimedia or video files available on server in form of numerous versions, an iPhone end user can not stream a better high quality version of the multimedia or video than iPad end user watching over 3G network.

2) Achieve more audiences - Transfer protocols are not supported for video delivery contents but firewall and routers settings are supported for video delivery with Hyper text transport protocol (HTTP) that's why viewers can access video easily.

3) Profit on bits transfer - With the help of HTTP live streaming , User can download a couple of segments of multimedia or video at time, that time user have to pay only transferred stream data . In addition, HTTP bits are cacheable by browsers or CDN and throughout network system.

4) Protected video clip information- The HTTP Live Streaming(HLS) requirements have conditions to make sure protection of the stream data, so it is fantastic information for Tv-stations or marketers for those users used to certified content stream. Using AES-128, the complete HTTP Live Streaming (HLS) stream is protected over network infrastructure.

Figure 2 and Explanation shows my practical work for mobile video streaming on Cloud with streaming server by using Amazon CloudFront services which have lots of components which are playing key role.

5.2.1. Explanation of R&D work

The characteristics of Adobe Flash Media Server (AFMS) version 4.5 can be utilized by Amazon Web Services (AWS) with live multimedia or video streaming, a sequence of HTTP requests from the end user devices deliver live video stream which is handled by manifest data files. By using Flash Media Server(FMS), end user can use two kind of HTTP file models, one is HTTP live streaming (HLS) for Apple company Products (Ios,iPad,iPhone etc) and second is HTTP Dynamic Streaming (HDS) for Flash type of programs or applications. By utilizing the Flash Media Live Encoder, viewer can get good quality of media streaming for different platforms or operating systems like windows and Macontish OS.

On-demand Real Time Messaging Protocol (RTMP) streaming from FMS would be assisted by CloudFront Information delivery program. It provides the flexible low cost Content Delivery Network (CDN) alternative for multimedia based organizations.AWS charge the cost only when user takes or uses the AWS services.

In this approach, make useful actions to set up CloudFront streaming:

    a. For live content delivery, create Simple Storage Service account known as S3.
    b. Create a "bucket" in S3 to store media files.
    c. Shift content to S3 bucket and set its permissions to allow public access.
    d. Set up a CloudFront streaming distribution that point at S3 storage bucket.
    e. Now you are ready to stream.

CloudFront provides the on demand multimedia streaming services with the help of Real Time Messaging Protocol by using Adobe Flash Media Server

The following versions of the RTMP protocol facilitated by CloudFront:

    a. RTMP—Adobe's Real-Time Message Protocol
    b. RTMPT—Adobe streaming tunneled over HTTP
    c. RTMPE—Adobe encrypted over HTTP
    d. RTMPTE—Adobe encrypted tunneled over HTTP

RTMPE is most secured protocol than RTMP.

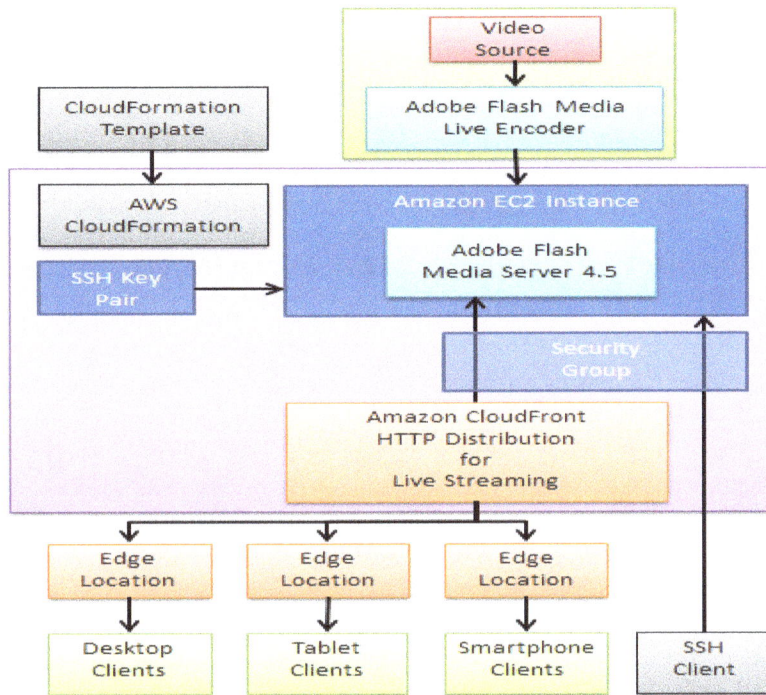

Figure 2. CloudFront Live streaming architecture

Many reputed IT companies are using HTTP Live streaming service to enhance the streaming power in their mobile domain infrastructure.

- Adobe Systems for Adobe Flash Media Server product.
- Livestation for multimedia channels France 24, RT, and Al Jazeera English.
- Microsoft in IIS Media Services 4.5.
- Google in Android 3.0 Honeycomb.
- HP in webOS 3.0.5.
- FFMPEG added HTTP Live Streaming and Encoding support for various mobile devices [11] [12].
- Zencoder Cloud based Encoding API added HTTP Live streaming support for iPad, iPod Touch and Apple TV [13].

6. CONCLUSION

In this paper, we have discussed firstly in audio/video basics which deliver video on network infrastructure with required bandwidth and codecs then after we discussed about the video streaming architecture that develop streaming servers which are responsible for downloading, uploading and adapting the video stream content in 3G or others networks.

For streaming the multimedia file over network, video compression techniques are major issue to encode the different types of audio/video file for different mobile devices. Compression can be performed by FFMPEG Encoding; Zencoder cloud based Encoding API which provides lots of Encoding techniques which are solution for the cloud based environments.

Then we presented the main issue of video streaming techniques for streaming the video over the internet or cloud based network for iPhone, Android, Window phone and Smartphone. Apple Company provides the solution for video streaming in terms of HTTP Live streaming which are accepted by many reputed companies for mobile devices for video streaming purpose for future perspective by using RTMP family protocols.

I believe that a lot of effort should be done in this paper to propose efficient and viable solution for mobile video streaming in cloud based environment.

ACKNOWLEDGEMENTS

There are some skilled personality's roles involved in this paper to achieve the goal. I show my honest thanks and appreciation to them who helped me lot.
I would like to thank my R&D team members of Pariksha Labs Pvt. Ltd, Gurgaon, India.

REFERENCES

[1]   Considerations for Creating Streamed Video Content over 3G-324M Mobile Networks, White paper. www.dialogic.com.

[2]   Prof. Nitin. R. Talhar, Prof. Mrs. K. S. Thakare "Real-time and Object-based Video Streaming Techniques with Application to Communication System", Proc .of CSIT vol.1 (2011) © (2011) IACSIT Press, Singapore.

[3]   Hatem BETTAHAR, "Tutorial on Multicast Video Streaming Techniques", SETIT 2005, 3rd International Conference: Sciences of Electronic, Technologies of Information and Telecommunications March 27- 31, 2005 –TUNISIA.

[4]   Mamoona Asghar, Saima Sadaf , Kamran Eidi, Asia Naseem, Shahid Naweed "SVS - A Secure Scheme for Video Streaming Using SRTP AES and DH", European Journal of Scientific Research ISSN 1450-216X Vol.40 No.2 (2010), pp.177-188 © EuroJournals Publishing, Inc. 2010.

[5]   Jian Zhou, "New Techniques for Streaming MPEG Video over the Internet", UMI Microform 3111144, Copyright 2004 by ProQuest Information and Learning Company, pp. 11-26.

[6]   M D Walker, M Nilsson, T Jebb and R Turnbull "Mobile video-streaming", BT Technology Journal-Vol 21 No 3- July 2003.

[7]   Jianyu Dong "Efficient and Effective Streaming Technologies for 3-D Wavelet Compressed Video", the Ohio State University 2002.

[8]   Delivering content to Apple iPhone, iPod Touch and iPad using RealNetworks Helix Solutions ©2010 RealNetworks, http://www.real.com.

[9]   Streaming media from Wikipedia available at http://en.wikipedia.org/wiki/Streaming_media.

[10]   Amazon CloudFront available at http://aws.amazon.com/cloudfront/.

[11]   FFMPEG Streaming Guide available at http://ffmpeg.org/trac/ffmpeg/wiki/StreamingGuide.

[12]   FFMPEG x264 Encoding Guide, http://ffmpeg.org/trac/ffmpeg/wiki/x264EncodingGuide.

[13]   Zencoder cloud based iOS/Mobile Encoding, https://app.zencoder.com/docs/guides/encoding-settings/ios-and-mobile.

3

IMPACT OF LIMITED FEEDBACK ON MIMO-OFDM SYSTEMS USING JOINT BEAMFORMING

NAJOUA ACHOURA[1] AND RIDHA BOUALLEGUE[2]

[1]Department National Engineering School of Tunis, Tunisia
najoua.achoura@gmail.com
[2] SUP'COM, 6'Tel Laboratory,Tunisia
ridha.bouallegue@supcom.rnu.tn

ABSTRACT

In multi input multi output antenna systems, beamforming is a technique for guarding against the negative effects of fading. However, this technique requires the transmitter to have perfect knowledge of the channel which is often not available a priori. A solution to overcome this problem is to design the beamforming vector using a limited number of feedback bits sent from the receiver to the transmitter. In the case of limited feedback, the beamforming vector is limited to lie in a codebook that is known to both the transmitter and receiver.When the feedback is strictly limited, important issues are how to quantize the information needed at the transmitter and how much improvement in associated performance can be obtained as a function of the amount of feedback available.In this paper channel quantization schema using simple approach to codebook design (random vector quantization)is illustrated. Performance results show that even with a few bits of feedback, performance can be close to that with perfect channel knowledge at the transmitter.

KEYWORDS

MIMO, OFDM, Linear precoding, CSIT, Random Vector Quantization.

1. INTRODUCTION

Multiple-input multiple-output (MIMO) systems, which use multiple antennas at both transmitter and receiver, provide spatial diversity that can be used to mitigate signal-level fluctuations in fading channels [2].When the channel is unknown to the transmitter, diversity can be obtained by using space-time codes [3],[2].When channel state information (CSI) is available at the transmitter, however, diversity can be obtained using a simple approach known as transmit beamforming and receive combining. Compared with space-time codes, beamforming achieves the same diversity order as well as additional array gain, thus it can significantly improve system performance. This approach, however, requires knowledge of the transmit beamforming vector at the transmitter. In practice, CSIT must be provided to the BS by some form of feedback.

CSIT feedback schemes are a very active area of research (see for example [13] and the special issue [14] for a fairly complete list of references). In brief, we may identify three broad families: 1) open-loop schemes based on channel reciprocity and uplink training symbols, applicable to Time-Division Duplexing (TDD); 2) closed-loop schemes based on feeding back the unquantized channel coefficients (analog feedback); 3) closed-loop schemes based on explicit quantization of the channel vectors and on feeding back quantization bits, suitably channel-encoded (digital feedback).

When the uplink and downlink channels are not reciprocal (as in a frequency division duplexing system), this necessitates that the receiver informs the transmitter about the desired transmit beamforming vector through a feedback control channel. The beamforming techniques proposed for narrowband channels can be easily extended to frequency selective channels by employing orthogonal frequency division multiplexing (OFDM). The combination of MIMO and OFDM (MIMO-OFDM), converts a broadband MIMO channel into a set of parallel narrowband MIMO channels, one for each subcarrier [10]. Transmit beamforming can be performed independently for each subcarrier of MIMO-OFDM. In non-reciprocal channels, this means requires that the MIMO-OFDM receiver calculates and sends back to the transmitter the optimal beamforming vector for every subcarrier. Practically, the feedback rate can be managed by using limited feedback techniques where the beamforming vectors are quantized using a beamforming codebook designed for narrowband MIMO channels [17]. For MIMO-OFDM's structures different approaches are possible. In this paper, joint beamforming, that consists in the extension to the transmitter side of the classical receive beamforming, is used. We focused this analyse on the impact of limited CSI on the transmitter on the performance of such system. This paper is organized as follows. In Section 2 we present the basic system model in brief. Section 3 introduce channel quantization limited feedback model, random vector quantization is presented in section 4. Finally in Section 5 some simulation results and conclusions are presented.

2. SYSTEM MODEL

In adaptive Beamforming, an array of antennas is exploited to reach maximum reception in a specified direction: the idea is to estimate the signal arrival from the desired direction (in the presence of noise) while signals of the same frequency from other directions are not accepted [2]. This can be achieved by varying the weights of each of the antennas used in the array. This spatial separation aims to separate the desired signal from the interfering signals. In adaptive beamforming the optimum weights are computed using complex algorithms based upon different criteria.

The communication over a frequency selective MIMO channel with N_T transmits and N_R receive antenna can be represented in multi-carrier fashion as:

$$y_k = H_k s_k + n_k \quad 1 \leq k \leq N \qquad (1)$$

Where k denotes the carrier index , N is the number of carriers, $s_k \in \square^{n_T \times 1}$ is the transmitted vector , $H_k \in \square^{n_R \times n_T}$ is the channel matrix, $y_k \in \square^{n_R \times 1}$ is the received signal vector, and $n_k \in \square^{n_R \times 1}$ is a zero-mean circularly symmetric complex Gaussian noise vector with arbitrary covariance matrix R_k. Since transmit beamforming is used at each carrier, the transmitted signal is:

$$s_k = b_k x_k \quad 1 \leq k \leq N \qquad (2)$$

Where b_k is the transmit beamvector and x_k is the transmitted symbol at the kth carrier [2].The receiver also uses beamforming:

$$\hat{x}_k = a_k^H y_k \quad 1 \leq k \leq N \qquad (3)$$

Where $a_k \in \square^{n_R \times 1}$ is the receive vector and x_k is estimated symbol at the kth carrier. The transmitted is constrained in its average transmit power as:

$$\sum_{k=1}^{N} E\left\{\|s_k\|^2\right\} = \sum_{k=1}^{N} \|b_k\|^2 \leq P_t \qquad (4)$$

where P_T is the power per block-transmission. Employing limited feedback in coherent MIMO-OFDM communication systems requires cooperation between the transmitter and receiver. At the reception, the estimate of the forward link channel matrix H is used to design feedback that the transmitter uses to adapt the transmitted signal to the channel.

There are two approaches to design feedback: quantizing the channel or quantizing properties of the transmitted signal. For most closed-loop schemes, either method can be employed. It will be apparent, however, that channel quantization offers an intuitively simple approach to closed-loop MIMO, but lacks the performance of more specialized feedback methods

3. CHANNEL QUANTIZATION

The basic idea behind closed-loop MIMO is to adapt the transmitted signal to the channel. One approach to limited feedback is to employ channel quantization, which is illustrated in Fig. 1. This problem is reformulated as a Vector Quantization problem (VQ) by stacking the columns of the channel matrix H into a $Mr \times Mt$ dimensional complex vector h_{vec}. The vector h_{vec} is then quantized using a VQ algorithm.

A vector quantization works by mapping a complex valued vector into one of a finite number of vector realizations. The mapping is usually designed to minimize some sort of distortion function such as the average mean squared error (MSE) between the input vector and the quantized vector.

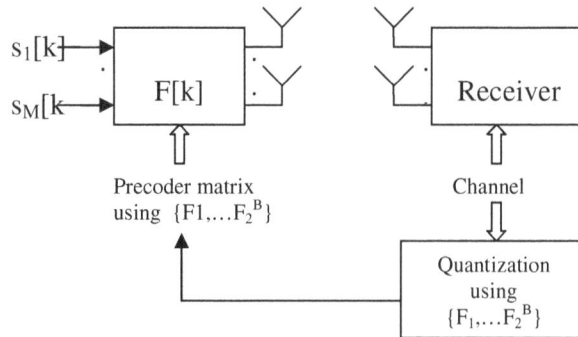

Figure 1. Limited feedback linear precoded MIMO system

Sending a quantized version of the forward link channel from receiver to transmitter gives the transmitter more flexibility to choose among different space-time signaling techniques.
Intuitively, one might expect that a random selection of matrices in the codebook F is likely to result in a large subspace distance between any pair of matrices in the codebook. This intuition is valid for a large number of antennas Mt, and is related to the fact that two vectors with i.i.d. components become orthogonal (with probability one) as the length becomes large. In the case of a random MIMO channel with i.i.d. components, the columns of the optimal precoding matrix are eigenvectors of the channel covariance matrix, which are isotropically distributed. These considerations motivated the Random vector quantization (RVQ) scheme proposed in [7], in which the elements of the codebook F are independently chosen random unitary matrices (i.e., Fk*Fk = I for each k).When used for beamforming in a MISO channel, RVQ is asymptotically optimal in the sense that it achieves the maximum rate over any codebook.

Furthermore, the achievable rate can be computed for both MISO and MIMO channels [7]. Here asymptotic means for a large system in which the number of antennas Mr and Mt each go to infinity with fixed ratio (or in the MISO case Mt goes to infinity), while also fixing B/MtM, the number of feedback bits per dimension.

3.1. Digital Channel Feedback Model

At the beginning, each receiver quantizes its channel to B bits and feeds back the bits perfectly to the access point. When each mobile has a single antenna (N=1), vector quantization is performed using a codebook C that consists of 2^B M-dimensional unit norm vectors $C \ \{w_1, \cdots w_{2^B}\}$. Each receiver quantizes its channel vector to the quantization vector that forms the minimum angle to it [4] [5]. Thus, user ι quantizes its channel to $\hat{h}_i$, chosen according to:

$$\hat{h}_i = \arg \min_{w = w_1, \cdots, w_{2^B}} \sin^2(\angle(h_i, w))$$

and feeds the quantization index back to the transmitter.

In this work we use random vector quantization (RVQ), in which each of the 2^B quantization vectors is independently chosen from the isotropic distribution on the M-dimensional unit sphere [6]. Each receiver is assumed to use a different and independently generated codebook, and we analyze performance averaged over the distribution of random codebooks.

When $N > 1$, the quantization codebook consists of matrices and the distance metric can be appropriately defined. Furthermore, random quantization corresponds to choosing the quantization matrices independently from the set of all unitary matrices.

3.2. Linear Precoding

After receiving the quantization indices from each of the mobiles, the AP uses linear precoding to transmit data to the mobiles. When N = 1, we consider the simple strategy of zero-forcing beamforming (ZFBF). Since the transmitter does not have perfect CSI, ZFBF is performed based on the quantizations instead of the channel realizations. When ZFBF is used, the transmit vector is defined as $x = \sum_{i=1}^{M} v_i s_i$ where each σ_ι is a scalar (chosen complex Gaussian with power P/M) intended for the ι-th receiver, and and $v_\iota \ 2 \ C_M$ is the beamforming vector for the ι-th receiver. The beamforming vectors $v_1, \ldots, v_M$ are chosen as the normalized rows of the matrix $[\hat{h}_i, \ldots, \hat{h}_M]^{-1}$, and thus they satisfy $\|v_j\| = 1$ for all $\hat{h}_i^H v_j = 0$ for all $\varphi \ 6 = \iota$. The resulting SINR at the ι-th mobile is [8]:

$$SINR_i = \frac{\frac{P}{M}|h_i^H v_i|^2}{1 + \sum_{j \neq i} \frac{P}{M}|h_i^H v_j|^2}$$

The achievable long-term average rate is the expectation of log $(1+\Sigma INP\iota)$ over the distribution of the fading and RVQ. When $N > 1$, ZFBF can be generalized to block diagonalization [9].

4. RANDOM VECTOR QUANTIZATION

In [12], authors analyzed, with limited channel knowledge at the transmitter, the channel capacity with perfect channel knowledge at the receiver,. Specifically, the optimal beamformer is quantized at the receiver, and the quantized version is relayed back to the transmitter. Given the quantization codebook $C = \{w_1, \cdots w_{2^B}\}$, which is also known a priori at the transmitter, and the channel H, the receiver selects the quantized beamforming vector to maximize the instantaneous rate, [11]

$$w(H) = \arg\max_{v_j \in V} \left\{ \log(1 + \rho \|Hw_j\|^2) \right\}$$

where $\rho = 1/\sigma_n^2$ is the background signal-to-noise ratio (SNR). The (uncoded) index for the rate-maximizing beamforming vector is relayed to the transmitter via an error-free feedback link. The capacity depends on the beamforming codebook V and B. With unlimited feedback (B→ ∞) the w(H) that maximizes the capacity is the eigenvector of H*H, which corresponds to the maximum eigenvalue.

We will assume that the codebook vectors are independent and isotropically distributed over the unit sphere. It is shown in [12], that this RVQ scheme is optimal (i.e., maximizes the achievable rate) in the large system limit in which (B,Nt,Nr)→∞ with fixed normalized feedback B = B/Nt and ¯Nr = Nr/Nt. (For the MISO channel Nr = 1). Furthermore, the corresponding capacity grows as log(ρNt), which is the same order-growth as with perfect channel knowledge at the transmitter. Although strictly speaking, RVQ is suboptimal for a finite size system, numerical results indicate that the average performance is often indistinguishable from the performance with optimized codebooks [12], [14].

5. SIMULATIONS RESULTS:

In this section, we evaluate the impact of working with limited feedback; witch is more practical, on system's performance, especially on the bit error rate. We consider only the case of one user (mono-user system). That means there is no need to user selection algorithms and the is no interference .Fig 2 shows the system's performance when there is two antenna in the transmitter and only one antenna on the reception (MISO system).

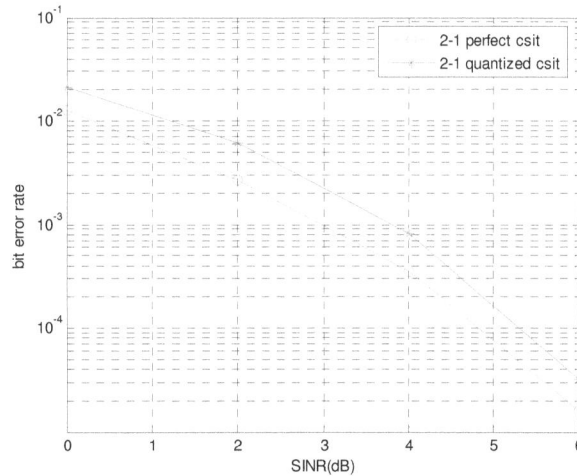

Figure 2. Nt=2, Nr=1 (with perfect channel knowledge)

Results show that for Nt = 2 and Nr=1, the Bit error rate with perfect channel knowledge at both the transmitter and receiver is larger than the rate with random vector quantified feedback. And this is perfectly expected so the blue curve is considered as the ideal case

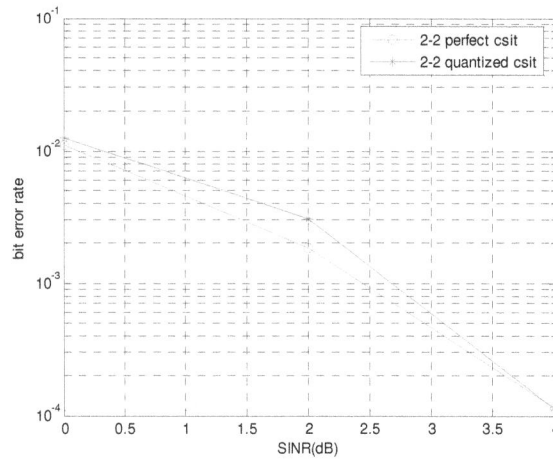

Figure 3. Nt=2, Nr=2

In Figure 3, we consider MIMO system with two antennas both in the transmitter and the receiver, using the same parameters. Results are better here so red curve witch represents quantified feedback is more close to the ideal case.

In other hand, we consider the case where we use channel estimation in figure 4, we consider the case when all training sequences are dedicated for estimation i.e. there is no data blocs. We show estimation's result for orthogonal phase shift sequences,

In fact we can easily remark degradation of performance between the two figures, this of course can be explained by the addition of two types of error: error due to channel estimation transmitted by feedback and the second error called quantization error.

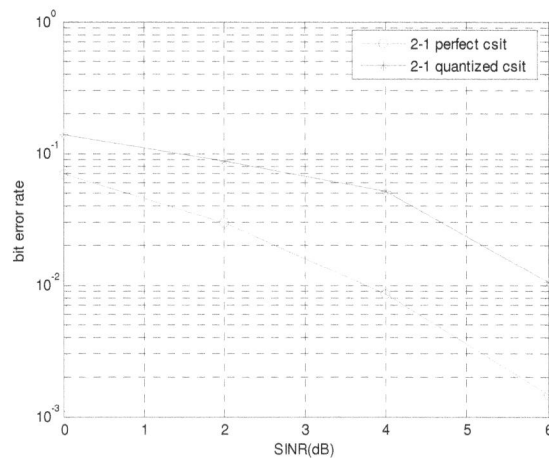

Figure 4. Nt=2, Nr=1 (with channel estimation)

3. CONCLUSIONS

This paper outlines a general framework for enabling limited feedback in closed-loop MIMO-OFDM systems. We review the application of limited feedback to MIMO communication. Numerical examples illustrate that relatively little feedback can provide substantial performance improvements.

Channel estimation error and channel evolution will definitely compromise expected performance improvements, but simulations and experimental results are required to determine how "recent" the feedback bits must be to maintain satisfactory performance.

More work is also needed in the area of limited feedback applications in MIMO-OFDM systems. In fact, a more practical technique is to feed back information on a select subset of tones and then use interpolation techniques. Other applications of limited feedback such as for multi-user MIMO channels are promising areas for investigation.

REFERENCES

[1]   D. J. Love, R. Heath., W. Santipach, and M. L. Honig, "What is the value of limited feedback for MIMO channels" IEEE Commun. Mag., vol. 42, no. 10, pp. 54–59, Oct. 2004

[2]   S. M. Alamouti, "A simple transmit diversity technique for wireless communications," IEEE J. Select. Areas Commun., vol. 16, pp. 1451–1458, Oct. 1998.

[3]   A. Iserte, A. Perez-Neira, D. Palomar, and M. Lagunas, "Power allocation techniques for joint beamforming in OFDM-MIMO channels," Proceedings EUSIPCO 2002, (France), September 2002.

[4]   D.J Love, R. Heath, and T. Strohmer, "Grassmannian beamforming for mimo wireless systems," IEEE Trans. Inform. Theory, vol. 49, no. 10, Oct. 2003.

[5]   C. Au-Yeung and D. Love, "On the Performance of Random Vector Quantization Limited Feedback Beamforming in a MISO System," IEEE Transactions on Wireless Communications, vol. 6, pp. 458-462, Feb. 2007.

[6]   W. Santipach and M.L. Honig, "Asymptotic capacity of beamforming with limited feedback," in Proceedings of Int. Symp. Inform. Theory, July 2004, p. 290.

[7]   R.Francisco and T. M. Slock, "An Optimized Unitary Beamforming Technique for MIMO Broadcast Channels », IEEE transaction on wireless communications, vol9, no. 3, march 2010

[8]   N. Ravindran, and N. Jindal, "Multi-User Diversity vs. Accurate Channel State Information in MIMO Downlink Channels", IEEE Trans. Communications, July 2009

[9]   N. Jindal, "MIMO Broadcast Channels with Digital Channel Feedback ", Asilomar Conference on Signals, Systems, and Computers, Asilomar, CA, Oct. 2006.

[10]   H. Bolcskei, M. Borgmann, and A. Paulraj, "Impact of the propagation environment on the performance of space-frequency coded MIMO-OFDM," IEEE J Select. Areas Commun., vol. 21, pp. 427–439, Apr. 2003.

[11]   W. Santipach, M. Honig "Optimization of Training and Feedback Overhead for Beamforming over Block Fading Channels" IEEE Trans. Info. Theory, August 2009

[12]   W. Santipach and M. L. Honig, "Capacity of multiple-antenna fading channel with quantized precoding matrix," IEEE Trans. Info. Theory, vol. 55, no. 3, pp. 1218–1234, Mar. 2009.

[13]   G. Caire, N. Jindal, and N. Ravindran, "Multiuser MIMO Downlink Made Practical: Achievable Rates with Simple Channel State Estimation and Feedback Schemes," *Submitted to IEEE Trans. Information Theory*, Nov. 2007, Arxiv preprint cs.IT/0711.2642v1.

[14]   IEEE Journal on Selected Areas on Communications (Special Issue on Limited Feedback), Nov. 2007.

4

MULTILEVEL ADDRESS REORGANIZATION TYPE2 IN WIRELESS PERSONAL AREA NETWORK

Debabrato Giri[1] and Uttam Kumar Roy[2]

[1]Department of Information Technology, Jadavpur University, Kolkata, India
debabrato.giri@tcs.com
[2] Department of Information Technology, Jadavpur University, Kolkata, India
u_roy@it.jusl.ac.in

ABSTRACT

The Standard for Wireless personal area network (WPAN) has been given by IEEE 802.15.4-2003 these networks are consists of Low rate, Low powered, Low memory devices. ZigBee Alliance has provided Network layer specification and Physical layer (PHY) and Medium access control (MAC) specification has been given by IEEE. In general the networks are of two kind Tree/Mess. In tree network no routing table is required for routing. After the great success in PAN this technique has also been tried to apply in business Network too. The main problem with this routing is that the maximum no of child (Router capable or end device) at any level and maximum network depth is fixed and this is done at the time of network formation. So the network can't grow beyond that max limit of breadth and width. We have addressed the network depth problem in our paper "Address Borrowing in Wireless Personal Area Network". Now in some other network configuration the maximum breadth of the network may be attained but the maximum depth of the network may not be attained (because of the asymmetric nature of the physical area) at that part and hence address lies unused. Here In this paper we have provided a unified address reorganizing scheme which can be easily applied to tree network so that the network can grow beyond the maximum no of child present at any level and overcome the address exhaustion problem by reorganizing address as per the requirement but without adding any extra overhead of having a routing table.

KEYWORDS

PAN; Mesh; Address Reorganizing; Routing; WPAN; Tree;

1. INTRODUCTION

In recent past there has been a steady rise in wireless networking. Wireless sensor/actuator networks ("sensornets") represents a new computing class consisting of large number of nodes which are often embedded in their operating environments distributed over wide geographical area often in remote and largely inaccessible regions. The node themselves ranges from tiny , resource-constrained devices called motes to PDA-class computing devices that are capable of sensing, computation, communication , and actuation. Sensornets allows us to instrument, observe, and respond to the physical world on scales of space and time previously impossible. Standards like IEEE 802.11[11] (Wi-Fi) and IEEE 802.15.1 [13] (12) have come into existence. IEEE 802.11 targets high data rate, mains powered, high cost and relatively long range applications. Bluetooth is one of the first standards designed for low range, low power devices. Because of its huge popularity now a day all most all the mobiles are coming with Bluetooth. But the main problem with this technology is the data can be transferred to single hop only. As a result more and more low-cost high-quality devices appear in the market; short-range low-rate wireless personal area networks are poised to take the world in a way observed never before which can transfer the data using multiple hops.

Wireless personal area networks (WPANs) are used to convey information over relatively short distances. Unlike wireless local area networks (WLANs), connections effected via WPANs involve little or no infrastructure. This feature allows small, power-efficient, inexpensive solutions to be implemented for a wide range of devices.

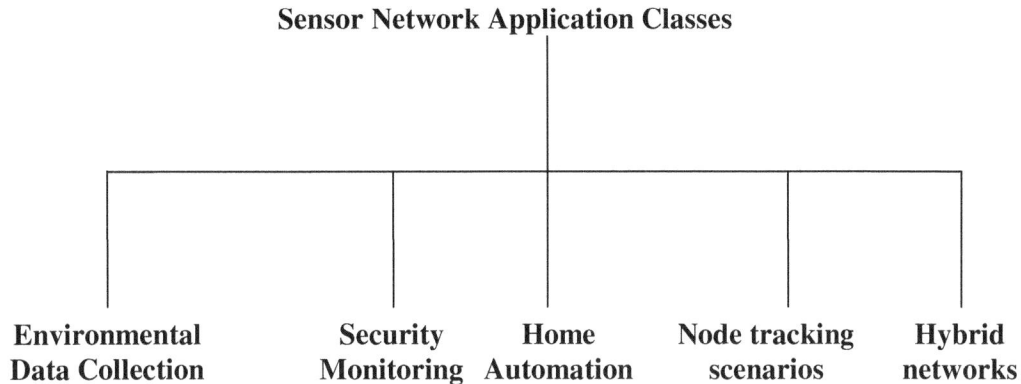

Sensor Network Application Classes

Environmental Data Collection	**Security Monitoring**	**Home Automation**	**Node tracking scenarios**	**Hybrid networks**

IEEE 802.15.4 [19] (henceforth referred to as 802.15.4) is a landmark in the attempt to bring ubiquitous networking [Figure 1] into our lives. Designed uniquely for energy-conscious low data rate appliances, it specifies the PHYsical (PHY) layer and MAC sub-layer of the protocol stack. The ZigBee Alliance [19] has defined the specification for the network (NWK), security and application profile layers for an 802.15.4-based system. Network layer supports three topologies Star, Tree and Mesh. The main advantage of tree address allocation is that it does not require any routing table to forward a message. In that simple mathematical equations are used for address assignment and routing.

With ZigBee devices on the horizon, ubiquitous networking looks elusive no more. It is not too distant a future one may chance to have one's home-appliances wedded together in a smart and cooperative network that allows them to talk to each other seamlessly. Sensors and actuators will communicate without barrier. They can be deployed pervasively in disaster-hit areas to monitor the situation to provide situational awareness and automatically take appropriate actions.
Some of its major application areas [Figure 1] are:

- Home automation
- Industrial control and monitoring
- Personal health-care
- Public safety including sensing, location determination and situational awareness at disaster sites
- Automotive sensing, such as tire pressure monitoring
- Precision agriculture such as the sensing of soil moisture, pesticide, herbicide, and pH levels.
- Mobile telecommunication such as peer-to-peer small data sharing, mobile commerce, mobile gaming, voice over ZigBee and chatting.

2. A QUICK TOUR OF 802.15.4/ZIGBEE

The following section gives a brief overview of ZigBee network formation and routing technique.

2.1. 802.15.4/ZigBee target applications

With ZigBee devices on the horizon, ubiquitous networking looks elusive no more. It is not too distant a future one may chance to have one's home-appliances wedded together in a smart and cooperative network that allows them to talk to each other seamlessly. Sensors and actuators will communicate without barrier. They can be deployed pervasively in disaster-hit areas to monitor the situation to provide situational awareness and automatically take appropriate actions.

Some of its major application areas [Figure 1] are:

- Home automation
- Industrial control and monitoring
- Personal health-care
- Public safety including sensing, location determination and situational awareness at disaster sites
- Automotive sensing, such as tire pressure monitoring
- Precision agriculture such as the sensing of soil moisture, pesticide, herbicide, and pH levels.
- Mobile telecommunication such as peer-to-peer small data sharing, mobile commerce, mobile gaming, voice over ZigBee and chatting.

Figure 1: An example of home automation

2.2. Highlights of 802.15.4 standard

This standard is a specification of the PHYsical layer (PHY) and Medium Access Control (MAC) sub-layer for low data rate wireless connectivity among relatively simple devices that consume minimal power and typically operate in a Personal Operating Space (POS) of 10 meters or less. The network can be a one-hop star, or, when lines of communication exceed 10 meters, a self-configuring, multi-hop ad-hoc network.

2.3. ZigBee Value Addition

The ZigBee Alliance defines the network (NWK), security, and application profile layers for the 802.15.4-based system. Two routing (Tree and Mesh routing) protocols have been defined. The algorithms are kept lightweight as devices are expected to be simple and have small memory.

Note that traditional table-driven ad hoc wireless routing protocols such as DSDV [2], CGSR, WRP, AODV [1, 3], require significant amount of memory for maintaining routing tables and hence are not suitable for these types of devices. Source-initiated routing such DSR [4], LMR, TORA, ABR, SSR may be considered as alternatives as they do not use any routing table. But, for a long network, incorporating routing information in the packet y the source is not practical due to the limitation on maximum packet size (16 bytes). Moreover, data rate is very low (e.g. not more that 100 packets throughout the day for a personal area network for home appliance) that requires a relatively light-weight routing algorithm which incurs an overhead as small as possible.

2.4. ZigBee Device Types

A device in a Zigbee network can be physically a Full Function Device (FFD) or a Reduced Function Device (RFD). An FFD typically has more resources than an RFD. Logically a Zigbee-device can be a

•ZigBee coordinator: 802.15.4 PAN coordinator for a ZigBee network—must be an FFD.

•ZigBee router: 802.15.4 FFD that is not the ZigBee coordinator but capable of being so and participates in mesh routing.

•ZigBee end-device: 802.15.4 RFD or FFD that is not a ZigBee coordinator..

2.5. Topologies Supported by Network Layer.

The ZigBee network (NWK) layer supports star, tree and mesh topologies. In a star topology, the network is controlled by one single device called ZigBee coordinator. The ZigBee coordinator is responsible for initiating and maintaining network. All other devices, known as end devices, directly communicate with the ZigBee coordinator. In mesh and tree topologies, the ZigBee coordinator is responsible for starting the network and for choosing certain key network parameters but the network may be extended through the use of ZigBee routers. In tree networks, routers move data and control messages through the network using a hierarchical routing strategy. Mesh networks allow full peer- to-peer communication.

2.6. Network Address Assignment

Network addresses are assigned using a distributed addressing scheme that is designed to provide a finite sub-block of network addresses to every potential parent. These addresses are unique within a particular network and are given by a parent to its children. The ZigBee coordinator determines the maximum number of children (*nwkMaxChildren*) that any device within its network is allowed. Of these children, a maximum of *nwkMaxRouters* can be router-capable devices while the rest will be reserved for end devices. But no hint is given how these parameters are determined. Every device has an associated depth, which indicates the minimum number of hops to reach ZigBee coordinator. The ZigBee coordinator itself has a depth of zero, while its children have a depth of one. Multi-hop networks have a maximum depth that is greater than one. The ZigBee coordinator also determines the maximum depth (*nwkMaxDepth*) of the network.

Given the following network parameters:

C_m = maximum number of children a ZigBee device may accept, *nwkMaxChildren*

L_m = maximum depth in the network, *nwkMaxDepth*

R_m = maximum number of router-capable-children a ZigBee device may accept, *nwkMaxRouters*

$E_m[=(C_m-R_m)]$ = maximum number of end-devices a ZigBee device may accept as children

We may compute the function, $C_{skip}(d)$, essentially the size of the address sub-block distributed by each parent at depth d to each of its router-capable child devices, as follows:

$$C_{skip}(d) = \begin{cases} 0 & : \ d = L_m \\ 1 + C_m.(L_m - d - 1), & : \ R_m = 1, \ d \neq L_m \\ \dfrac{1 + C_m - R_m - C_m.R_m^{L_m - d - 1}}{1 - R_m} & : \ R \neq 1, d \neq L_m \end{cases}$$

(1)

If a device has a $C_{skip}(d)$ value of zero, then it shall not be capable of accepting children and shall be treated as a ZigBee end device. A device that has a $C_{skip}(d)$ value greater than zero may accept child devices and may assign addresses to them differently depending on whether the child device is router-capable or not. Network addresses are assigned to router-capable child devices using the value of $C_{skip}(d)$ as an offset.

A router-capable device having address A_{parent} at depth **d** assigns addresses to its **nth** child **A$_n$** at depth **d+1** in the following way:

END-DEVICE CHILD:

$$A_n = A_{parent} + C_{skip}(d)R_m + n \ : 1 \leq n \leq E_m$$

(2)

ROUTER-CAPABLE CHILD:

$$A_n = A_{parent} + C_{skip}(d)(n-1) + 1 \qquad : 1 \leq n \leq R_m$$

(3)

The first router-capable child gets address $A_{parent}+1$ and subsequent router-capable children get addresses separated by $C_{skip}(d)$. End-devices get addresses starting from $A_{parent}+1+C_{skip}(d)*R_m$ separated by 1. Such example is shown in Figure. 3 and Figure 6.

2.7. Tree Routing Mechanism

For hierarchical routing, if the destination is a descendant of the device, the device shall route the frame to the appropriate child. Trivially, every other device is a descendant of the ZigBee Coordinator and no device is a descendant of any ZigBee end-device.

Refer to (1), (2) and (3), taking a routing decision is very simple. A target device with address D is a descendant of a ZigBee router with address A at depth d if

$$A < D < A + C_{skip}(d-1)$$

(4)

This follows because the parent P at depth *(d-1)* of the device X with address A, gives the address A to X and the address $A+C_{skip}(d-1)$ to the "closest" router-capable-sibling of X that is also a child of *P*. The address sub-block *[A+1, A+C$_{skip}$(d-1)-1]* is reserved for the descendants of *X*.

If the destination is a descendant of the receiving device, the address *N* of the next hop device is determined as:

$$N = \begin{cases} D, & \text{if } D > A + R_m \times C_{skip}(d) \quad \text{for end devices} \\ A+1+\left\lfloor \dfrac{D-(A+1)}{C_{skip}(d)} \right\rfloor \times C_{skip}(d), & \text{otherwise} \end{cases}$$

$$(5)$$

If equation (5) is not satisfied, next hop device is its parent. We look below at the derivation of formula (6):

The next-hop address is *N=A+1+kC$_{skip}$(d)*, i.e. the packet is to be routed to the router-capable-child with address *A+1+kC$_{skip}$(d)* if *A+1+kC$_{skip}$(d) ≤D< A+1+(k+1)C$_{skip}$(d)*

This is because the device with address *A+1+kC$_{skip}$(d)* gets address sub-block *[A+1+kC$_{skip}$(d),*

A+1+(k+1)C$_{skip}$(d)-1] i.e. $$N = A + 1 + kC_{skip}(d),$$ Where $$k \le \frac{D-A-1}{C_{skip}(d)} < k+1$$

$$\Rightarrow k = \left\lfloor \frac{D-A-1}{C_{skip}(d)} \right\rfloor$$

$$(6)$$

So, next hop router can be determined by using an equation and the complexity of this scheme is constant. Moreover, there is no routing table and that way searching procedure is completely eliminated. In spite of this, tree routing has several limitations as described in the next section.

2.8. Problem Definition

The fundamental query that arises regarding tree routing is "how to choose the values of C_m and R_m?". In many cases, before actually forming the network, we have very little or no idea about the following parameters:

- o   Number of end devices that will join to a router
- o   Number of routers that will join to a router
- o   Depth of the tree network

Also once the value is chosen the address distribution will be symmetric and it will not be able to support asymmetric structures required in mine field, glaciers sea bed, building premises etc.

Because an address sub-block cannot be shared between devices, it is possible that one parent exhausts its list of addresses while a second parent has addresses that go unused. A parent having no available addresses shall not permit a new device to join the network. In this situation, the new device shall find another parent. If no other parent is available within transmission range of the new device, the device shall be unable to join the network unless it is physically moved or there is some other change.

This routing technique uses a fixed address assignment i.e. the no of child are fixed at any level and hence the address range, it is possible to have a network in which the no of child per parent increases as the depth increases, in this kind of network a large range of address will remain unused at the top level

Finally, due to the fact that the tree, it is not dynamically balanced, the possibility exists that certain installation scenarios, such as long lines of devices, may exhaust the address capacity of the network long before the real capacity is reached.

Following section shows a typical network formation where for a part of the tree number of node at the higher layer is relatively high but no nodes are present at the lower level.

Let Cm=2, Lm=4, Rm=2

Depth in the Network	Offset Value [Cskip(d)]
0	15
1	7
2	3
3	1
4	0

Table 1 –Cskip

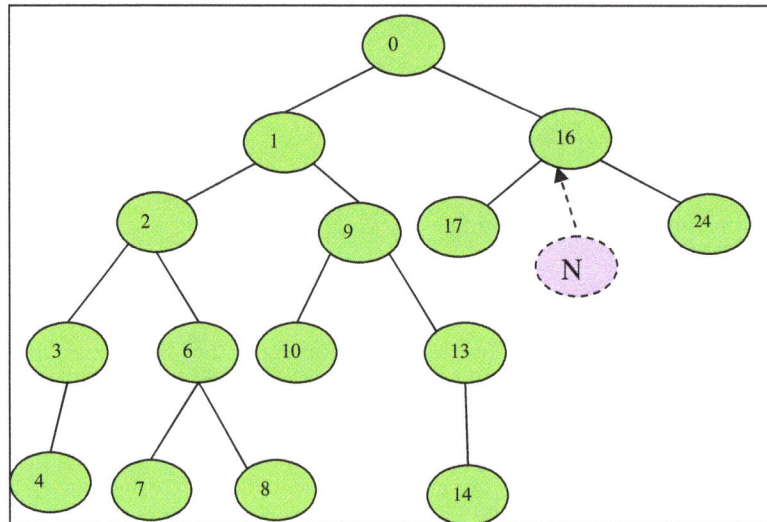

Figure 3: Limitation on maximum child for ZigBee tree networks

3. Related works

Wireless networks have rapidly gained popularity since their introduction in 1970s. However, an investigation into low-cost, low-rate, low-power PAN is relatively new.

In [6], we have provided a unified address borrowing scheme which can be easily applied to grow the network beyond 16 hops and overcome the address exhaustion problem by borrowing address. A routing algorithm based on mobile IP, is also proposed.

In [7], we extended the Tree routing proposed by ZigBee for the networks to be harsh and asymmetric.

4. Proposed Address Reorganization Algorithm

In this paper we have extended our solution of single level address re-organization to multiple levels with out any extra over head such as routing table. The next hop address is calculated using mathematical formula only .This algorithm will allow the formation of any asymmetric network as per the need and thus remove the limitation of symmetric address distribution and formation. In real world most of the networks area is asymmetric for example in a building the lower floors can have more rooms compared to upper floors and part of ground floor may have canteen/gym etc so does not require any network or a paddy field can have some irregular shape, proposed address reorganization technique is capable of handling all these and can be applied to all such asymmetric networks.

In this scheme a node will be allowed to join a network at a node even if it has reached maximum no child by Address reorganization. This scheme can be used in part of the network where the network wants to grow in breadth rather than in depth i.e. we are expecting that the depth will be less than Lm at that part of the network. In Figure 3 the maximum length of the path from root which goes via Node 16 is two i.e. the depth of the network at that part is 2(less than Lm) which suggests that at that part of the network the growth is around the breadth and not in depth . We can apply our algorithm at that part.

In proposed address reorganization scheme any parent can increase its no of child device by reorganizing its address by any level so it is much more flexible.

4.1. Overview of the Algorithm

The Node which has reached its maximum child and wants to expand its breadth will use the next level Cskip value while distributing the address to its child i.e. if the node is at level K it will use the Cskip value for K+1, its immediate child will use K+2 and so on till Lm-1.By doing this the Node has gained one level of address which it can use for adding new child.In Figure 3 if a new node wants to join the network at Node 16 then it will not be allowed as Node 16 has exhausted its address even though free address is available at other node such as Node 17 or Node 24.Our proposed algorithm could be used in such scenarios.

After address reorganizing the maximum no of child that can be added to that node will be Rm*Rm + Cm

For example if node 16 goes for address reorganization as it is at level 1 its original Cskip value is 7 but it will use its next level Cskip value i.e. level 2 which is 3, and the maximum no of child it can have will be 2*2+2=6 and the network depth at that part will be Lm-1=3.

So after address reorganization node 16 will be able to add 4 more children. Following Figure 4 shows the structure of the network after address reorganization

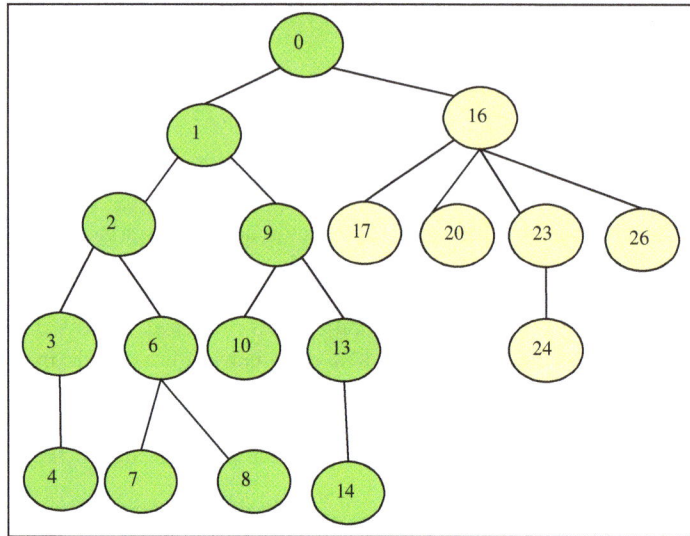

Figure 4 Address Reorganization

4.2. Required Data Structure

Transition table: This table will be maintained at the Node which has reorganized its address e.g. Node 16 in **Figure 4**. It will have 2 column Name and value.

Name	Value
Actual Cskip(Cskip)	7
Pseudo Cskip(Pcskip)	3
Actual Level(d)	1
Pseudo Level(Pd)	2

Table 2 –Transition Table

This table will have 4 rows and will be used by the address reorganizing Node while assigning address to its child devices.

In the above table Pseudo level (Pd)=d+1 and $P_{cskip}(d) = C_{skip}(d+1)$ **(7)**

All the child node of this will be at pseudo level Pd+1and there child at level Pd+2 and so on. That is the node which has re organized its address will physically be at level d but will virtually consider it to be at level d+1.

4.3. Address Distribution

In this variant if the node has reorganized its address by level 'v' the maximum no of child that the node can have is $(R_m^{v+1} + R_m)$

At any Node if it has not reorganized its address or any of its parents have not reorganized its address then the address assignment will follow equation 2 and 3 for Router and end device respectively.

If the Node has reorganized its address by level 'v' then the address assigned to its router capable device will be as follows:

For R^{th} routing capable child if R is $<= R_m^{v+1} + 1$

$$A_{R^{th}} = A_{parent} + P_{cskip}(d).(R^{th} - 1) + 1 \qquad (8)$$

Else if $R_m^{v+1} + 1 < R <= R_m^{v+1} + R_m \qquad (9)$

Then address is given by $A_{R^{th}} = A_{parent} + P_{cskip}(d)R_m^{v+1} + 1 + E \qquad (10)$

Where **G**$= (R_m + R_m^2 + ...R_m^v)$ & *for v >3*

$$E = (C_m - R_m + 1).(1 + R_m + R_m^2 + ...R_m^{v-1}).(R - R_m^{v+1} - 1) \qquad (11)$$

for v=1, $\qquad E = (C_m - R_m + 1).(R - R_m^{v+1} - 1) \qquad (12)$

for v=2, $\qquad E = (C_m - R_m + 1).(1 + R_m).(R - R_m^{v+1} - 1) \qquad (13)$

for v=3, $\qquad E = (C_m - R_m + 1).(1 + R_m + R_m^2).(R - R_m^{v+1} - 1) \qquad (14)$

These nodes will mark their relative position as 1. i.e. they are at 1 level below the node which has reorganized its address. The child node of this node will mark its relative position as two and so on. For example in **Figure-7** Node 42 and 51 are at relative level 1, node 43, 46, 52 and 55 are at relative level two. The Cskip value for a node at relative level 'e' *for'v'>3* is given by

$$R_{skip}(e) = (C_m - R_m + 1).(1 + R_m + R_m^2 + ...upto(v - e)term) \qquad (15)$$

for v-e=1, $\qquad R_{skip}(1) = (C_m - R_m + 1). \qquad (16)$

for v-e =2, $\qquad R_{skip}(1) = (C_m - R_m + 1).(1 + R_m) \qquad (17)$

for v-e =3, $\qquad R_{skip}(1) = (C_m - R_m + 1).(1 + R_m + R_m^2) \qquad (18)$

Network address to the end device is given in a sequential manner and the address given to nth end device is given by the following equation.

$$A_n = A_{Parent} + P_{cskip}(d).R_m^{v+1} + n + F \qquad (19)$$

$1 \le n \le Cm - Rm$ *And* A_{parent} *is address of parent And* $F = (C_m - R_m + 1).G$

All the nodes which are descendent of first R_m^{v+1} child of that reorganizing node will use the subsequent depth and Cskip values, and the node which are descendent of next Rm child will use Rskip value for address distribution. Please refer to **Figure-7** for network formation. In that node 1 has reorganized its address by level 2 so the no of router capable device that can be attached to it is $R_m^3 + R_m = 10$ among which 8 node have Pseudo Cskip here $P_{cskip} = 5$ and the other 2 child Node will have Relative Cskip here the value is $R_{skip}(1) = 3 \times 3 = 9$ according to **Equation 17** i.e. node & which is among 1st 8 child node will use $P_{cskip} = 5$ and node 42 will use $R_{skip}(1) = 3 \times 3 = 9$ for address distribution.

4.4. Additional Data Structure

In this type one additional register R_{pos} will be required for storing the relative position. This register's value will be set only if the Rth Node is a child of the address reorganizing node and if $R_m^{v+1} < R <= R_m^{v+1} + R_m$

or child of a node which satisfies the above two conditions for example in **Figure-7** node 42 has $R_{pos} = 1$ Node 43 has $R_{pos} = 2$

4.5. Routing in Reorganized Network

At any Node if it has not reorganized its address then the routing will follow the normal process i.e. it will follow equation 4,5and 6.

If the Node has reorganized its address then routing will be as follows:
If the destination is a descendant of the device, the device shall route the frame to the appropriate child. If the destination is not a descendant, the device shall route the frame to its parent. For a ZigBee router with address A at depth d, and pseudo depth Pd if the following logical expression is true, then a destination device with address D is a descendant:

$$A < D < A + C_{skip}(d-1) \tag{20}$$

If it is determined that the destination is a descendant of the receiving device then it is checked if the destination is a child end device using the formula

$$D > A + R_m^{v+1} . P_{cskip}(P_d) + G.(C_m - R_m + 1), \tag{21}$$

If it is a child end device then the address N of the next hop device is given by: N=D.

Otherwise,

If the destination address $A < D <= A + P_{cskip}(P_d).R_m^{v+1}$

then the next hop address is given by:

$$N = A + 1 + \left\lfloor \frac{D-(A+1)}{P_{cskip}(P_d)} \right\rfloor \times P_{cskip}(P_d) \tag{22}$$

else the next hop address is given by

$$N = Z + \left\lfloor \frac{D-Z}{R_{csik}} \right\rfloor \times (R_{cskip}) \tag{23}$$

Where $Z = A + 1 + R_m^{v+1} \times P_{cskip}(P_d)$ $\tag{23}$

All the nodes which are descendent of first R_m^{v+1} child of that reorganizing node shall use there pseudo depth and the node which are descendent of next Rm child shall use relative depth and relative Cskip value in place of Cskip in equation 4,5,6,7 for routing purpose, no other changes are needed .

A node will use R_{pos} register to decide if it needs to use P_{cskip} or $R_{skip}(e)$ for routing. If the R_{pos} register contains a not null value then it will use $R_{skip}(e)$ else it will use P_{cskip} if it is a descendent of address reorganized node and C_{skip} otherwise.

For example in **Figure-7** say node 71 wants to send some data to Node 34 it will go via the following path. **Figure -5** shows the flow chart for the routing.
$71 \Rightarrow 70 \Rightarrow 64 \Rightarrow 63 \Rightarrow 62 \Rightarrow 0 \Rightarrow 1 \Rightarrow 32 \Rightarrow 34$ .In Figure 8 the message flowing path has been highlighted in dark violet.

At Node 1 routing decision will be taken depending on **equation (20,21,22,23,24)** at all the other nodes routing will follow equation 3,4,5.
Again If Node 14 wants to send some data to node 114 it will
follow the following path.
$14 \Rightarrow 12 \Rightarrow 1 \Rightarrow 0 \Rightarrow 62 \Rightarrow 92 \Rightarrow 106 \Rightarrow 112 \Rightarrow 114$ In Figure 8 the message flowing path has been highlighted in dark blue.

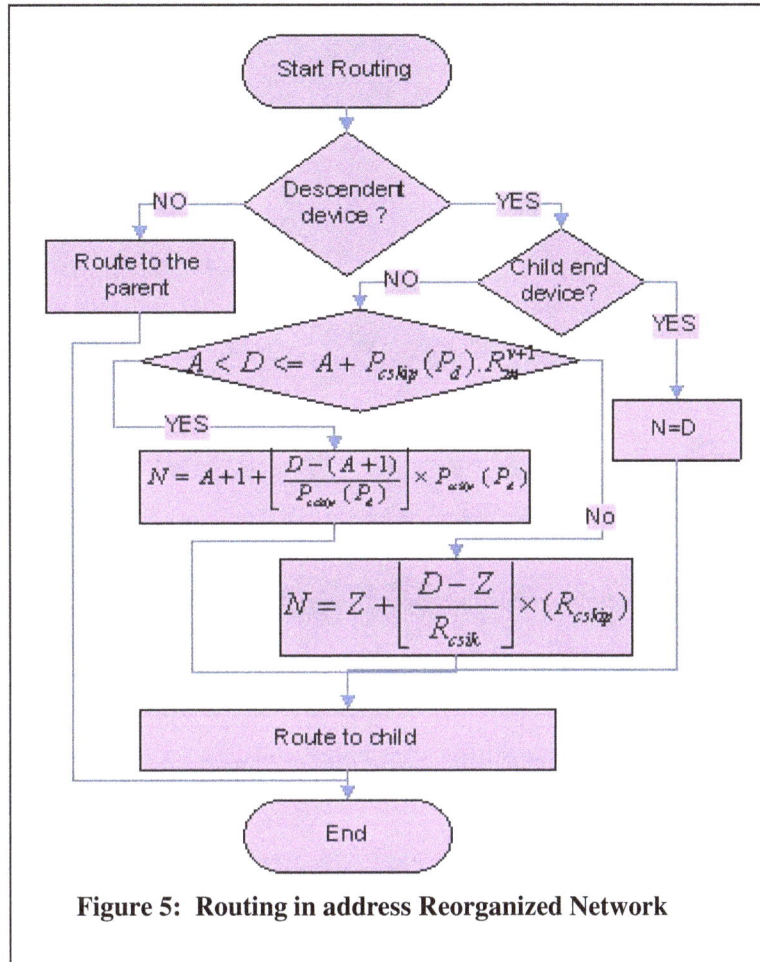

Figure 5: Routing in address Reorganized Network

$C_{skip}(0)$	$C_{skip}(1)$	$C_{skip}(2)$	$C_{skip}(3)$	$C_{skip}(4)$	$C_{skip}(5)$
61	29	13	5	1	0

Table 3: Cskip Value at various levels

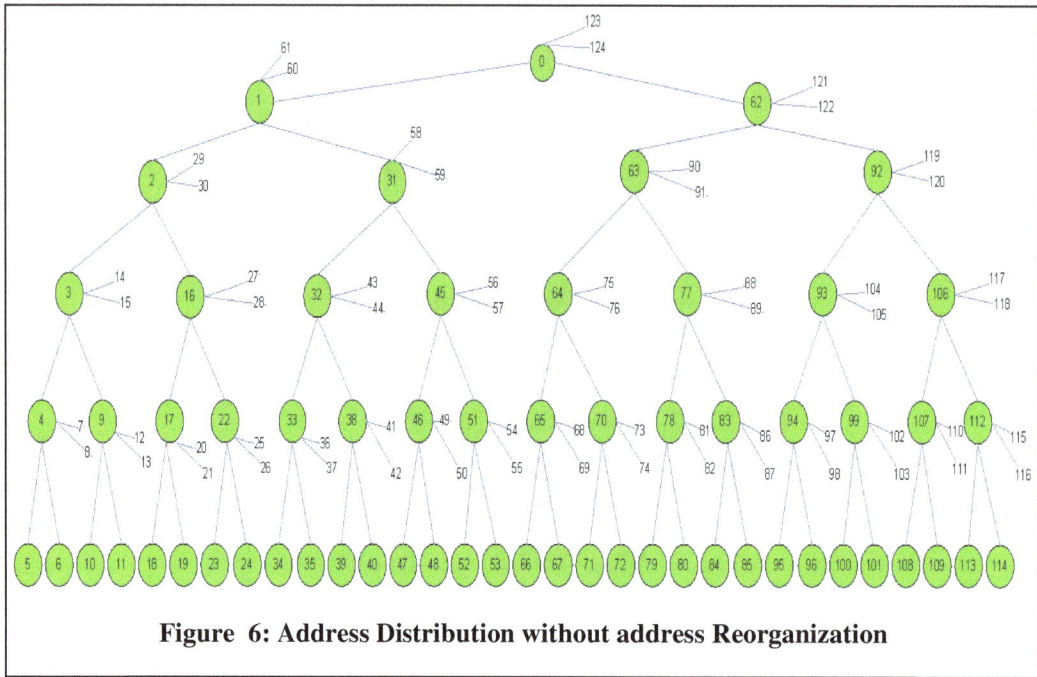

Figure 6: Address Distribution without address Reorganization

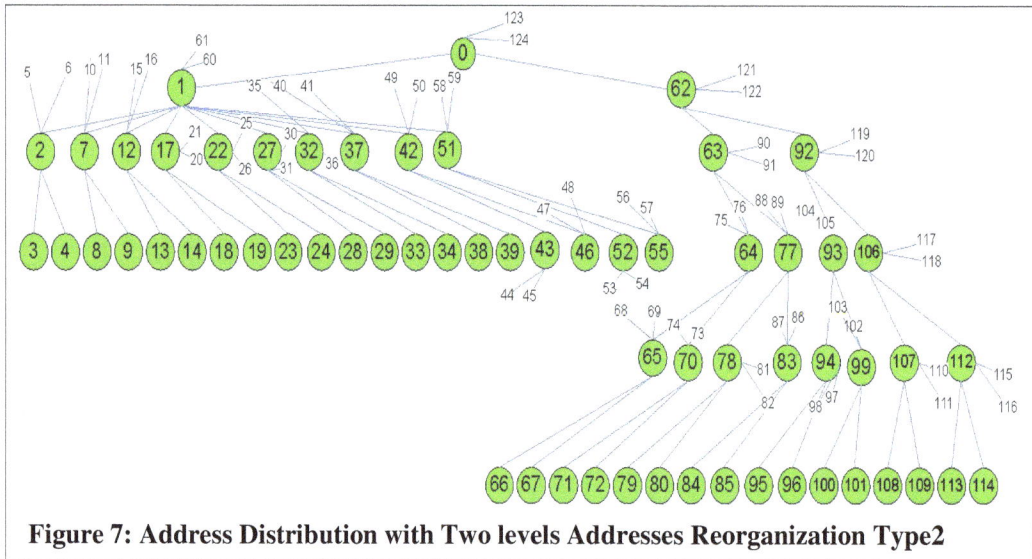

Figure 7: Address Distribution with Two levels Addresses Reorganization Type2

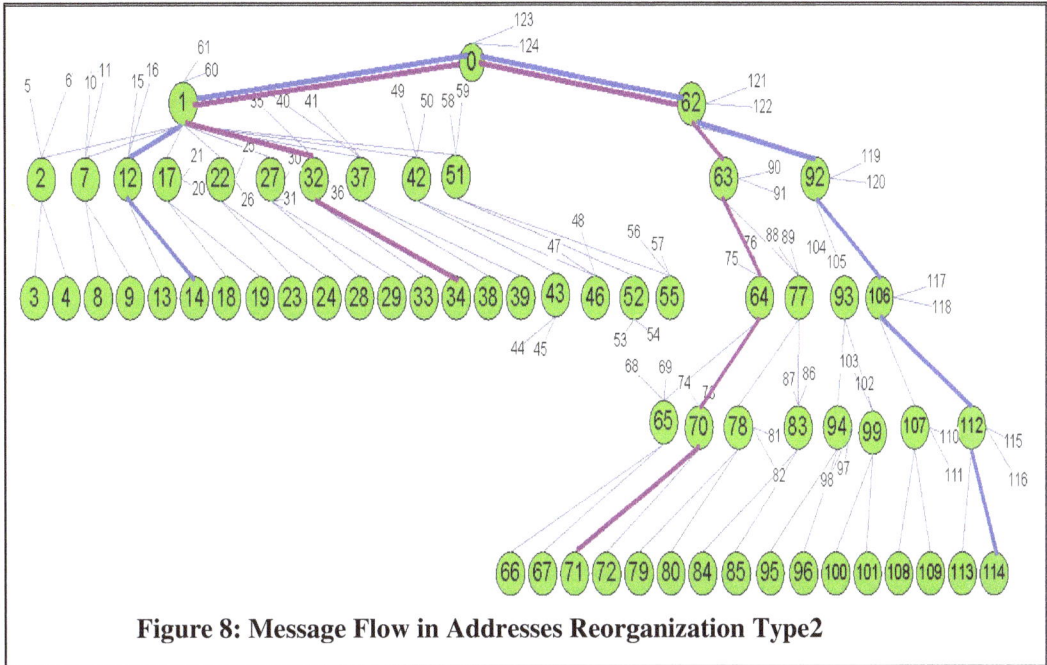

Figure 8: Message Flow in Addresses Reorganization Type2

4.6. Analysis of Overhead

In this address re-organized network if the node has reorganized its address by level 'v' the maximum no of router capable child that the node can have is $(R_m^{v+1} + R_m)$. For example in normal tree network (**Figure 6**) the **Node 1** can have at most 2 router capable device attached to it. But if the requirement is to add more router capable devices at this node then it can re-organize its address by any level .In **Figure 7** we have shown a re-organized network where **Node 1** has re-organized its address with 2 levels i.e. v=2.and Rm=2. So it can now have $(2^{2+1} + 2) = 10$ router capable device. Following table shows the relation ship between different address re-organization level(v) and the max no of router capable child that a re-organized node could have for a given Rm .For example if for a network Rm is chosen as 2.then If any node wishes to re-organize its address by level 1 it can now have 6 router capable child. If it has re-organized its address by level 3 then it can have 18 router capable child and so on.

No of re organization Level (v)	Max no of Router capable Child without address reorganization	Max No of Router capable Child with address reorganization
1	2	6
2	2	10
3	2	18
4	2	34
1	3	12
2	3	30
3	3	84
4	3	246

Table 4 : Max Router capable device.

No of Router capable device at any level v is R_m^v this entire node could re-organize its address by 1 to Lm –v-1 level so there will be (Lm-v-1) R_m^v pattern that could be formed at level v. So if we consider each router capable devices of all Lm levels then all together the no of patterns that could be formed will be.

$$\sum_{v=1}^{L_m-2} R_m^v (L_m - v - 1) \qquad (24)$$

Following table shows the relation ship between the no of level and the no of different pattern that could be supported for a given Rm, Lm. For example if Rm=2,Lm=3 then the different pattern formation additional to the normal network is 2.

Network Parameter	No of Different Structure Possible with address re-organization
RM=2,LM=3	2
RM=2,LM=4	8
RM=2,LM=5	22
RM=2,LM=6	52
RM=2,LM=7	114
RM=2,LM=8	240
RM=2,LM=9	494
RM=2,LM=10	1004
RM=2,LM=11	2026
RM=2,LM=12	4072
RM=2,LM=13	8166
RM=2,LM=14	16356
RM=2,LM=15	32738
RM=2,LM=16	85984

Table 5: Different Network Formation

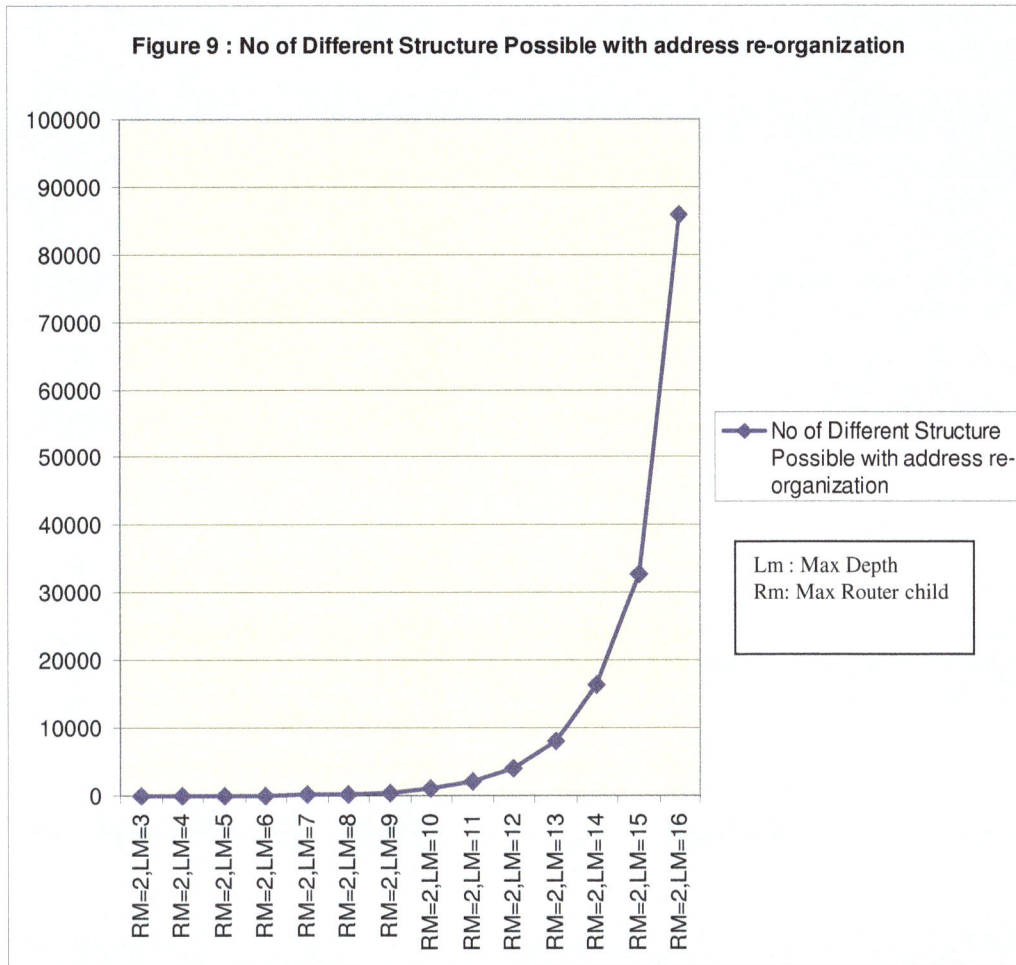

Figure 9 : No of Different Structure Possible with address re-organization

Lm : Max Depth
Rm: Max Router child

5. CONCLUSIONS

Sensornets are closely coupled to the physical world and can directly impact our Personal Operating Space (POS). One of the major advantages of IEEE802.15.4 is that it uses free 2.4 GHz ISM band at the Physical layer. Nodes in a sensornet are low cost and the performance of this architecture is already well-proven.

The wireless network with minimum data rate, reduced energy consumption and minimum cost is formed by IEEE 802.15.4 standard. The distinctive characteristics of the standard makes it more approving module for wireless sensor networks and remote monitoring applications. It offers a minimum power, economic and a consistent protocol for wireless connectivity among low-cost, permanent and moveable devices. These devices can figure out into a sensor network or wireless personal area network (WPAN). After a great success in personal operating space (POS), researchers are trying to use it in relatively broad areas such as industrial automation, tracking and monitoring systems, public telecom services etc. This is only possible if a large network can be formed. For this purpose it needs a suitable routing algorithm.

In this paper we have tried to mitigate the problem of symmetric structure and provided a unique technique of address reorganization which will help the network in a variety of different application scenarios staring from mine field to Glacier war field to building automation. The main advantage of this algorithm is that we can form many different network with the introduction of just two variable called Pseudo Cskip(Pcskip) and Pseudo Level(Pd) no other extra overhead such as routing table will not be required.

ACKNOWLEDGEMENTS

The authors would like to thank everyone, just everyone!

REFERENCES

[1] C. E. Perkins, E. Belding-Royer, and S. R. Das, "Ad hoc On-Demand Distance Vector (AODV) Routing", http://www.ietf.org/rfc/rfc3561.txt, July 2003. RFC 3561.[AODV_1]

[2] C. E. Perkins and P. Bhagwat, "Highly Dynamic Destination-Sequenced Distance-Vector Routing (DSDV) for Mobile Computers", Proceedings of ACM SIGCOMM, 1994.[C_E_DSDV]

[3] C. E. Perkins and Elizabeth Royer, "Ad-hoc On-Demand Distance Vector Routing", Proceedings of the 2nd IEEE Workshop on Mobile Computing Systems and Applications, New Orleans, LA, February 1999. [AODV_2]

[4] D. B. Johnson, D. B. Maltz, "Dynamic Source Routing in Ad-hoc Wireless Networks", Mobile Computing, T.   Imielinski, H. Korth, Eds. Kluwer Academic Publishers, 1996, ch. 5, pp. 153-181.[Johnson_DSR]

[5] D. Ganeshan, B. Krishnamachari, "Complex Behavior at Scale: An Experimental Study of Low-Power Wireless Sensor Networks", UCLA/CSD-TR 02-0013, UCLA Computer Science, 2002.[Low_Power]

[6] Debabrato Giri, Uttam Kumar Roy, "Address Borrowing In Wireless Personal Area Network", Proc. of IEEE International Anvanced Computing Conference, (IACC '09, March 6-7), Patiala, India, page no 1074-1079[ukr_borrowing]

[7] Debabrato Giri, Uttam Kumar Roy, *"Single Level Address Reorganization In Wireless Personal Area Network"*, 4th International Conference on Computers & Devices for Communication (CODEC-09), December 14-16, 2009, Calcutta University, India. [ukr_single]

[8] Ed Callaway, Paul Gorday, Lance Hester, Jose A. Gutierrez, Marco Naeve, Bob Heile and Venkat Bahl. "Home Networking with IEEE 802.15.4: A Developing Standard for Low-Rate Wireless Personal Area Networks",IEEE Communications Magazine August 2002.[Low_Rate]

[9] Elizabeth Royer and C-K Toh, "A Review of Current Routing Protocols for Ad-Hoc Mobile Wireless Networks", IEEE Personal Communications Magazine, April 1999.[Routing_Protocols]

[10] Gang Lu, Bhaskar Krishnamachari, Cauligi S. Raghavendra, "Performance Evaluation of the IEEE 802.15.4 MAC for Low-Rate Low-Power Wireless Networks", IEEE International Conference on Performance, Computing, and   Communications, 2004.[Performace]

[11] IEEE 802.15.11 Standard: "Wireless Local Area Networks, 1999". [WLAN]

[12] IEEE 802.15.1, Wireless Medium Access Control (MAC) and Physical layer (PHY) specifications for Wireless Personal Area Networks (WPANs) [Bluetooth]

[13] IEEE 802.15.4, Wireless Medium Access Control (MAC) and Physical Layer (PHY) Specifications for Low-Rate Wireless Personal Area Networks (WPANs) [WPAN]

[14] J. Heidemann, W. Ye and D. Estrin."An Energy-Efficient MAC Protocol for Wireless Sensor Networks", Proceedings of the 21st International Conference of the IEEE Computer and Communications Societies (INFOCOM 2002), New York, NY, June 2002.[Energy]

[15] Jianliang Zheng and Myung J. Lee. "A Comprehensive Performance Study of IEEE 802.15.4", http://www-ee.ccny.edu/zheng/pub, 2004. [Comprehensive]

[16] 14, C. Schurgers, S. Park and M. B. Srivastava, "Energy-Aware Wireless Microsensor Networks", IEEE Signal Processing Magazine, Volume: 19, Issue: 2, March 2002.[Aware]

[17] Shree Murthy, J. J. Garcia-Luna-Aceves, "An Efficient Routing Protocol for Wireless Networks", Mobile Networks and Applications, 1996.[Efficient]

[18] Uttam Kumar Roy, Debarshi Kumar Sanyal, Sudeepta Ray, "Analysis and Optimization of Routing Protocols in IEEE802.15.4" (Asian International Mobile Computing Conference (AMOC 2006), Jadavpur University, Kolkata, India [UKR_AMOC]

[19] ZigBee Alliance (ZigBee Document 053474r17) Draft Version 0.90: Network Specification, Jan 2008.[ZigBee]

[20] ZigBee Alliance (ZigBee Document 075307r07) Version 1.0: Telecom Applications Profile Specification, April 2010.[ZigBee Telecom]

[21] ZigBee Alliance (ZigBee Document 053516r12) Version 1.0: ZigBee Building Automation Application Profile, May 2011.[ZigBee Building Automation]

5

Multi-Hop Clustering Protocol using Gateway Nodes in Wireless Sensor Network

S. Taruna[1], Rekha Kumawat[2], G.N.Purohit[3]

[1]Banasthali University, Jaipur, Rajasthan
staruna71@yahoo.com
[2]Banasthali University, Jaipur, Rajasthan
er.rekha28@gmail.com
[3]Banasthali University, Jaipur, Rajasthan
gn_purohitjaipur@yahoo.com

ABSTRACT

Wireless sensor networks (WSNs) are composed of many homogeneous or heterogeneous sensor nodes with limited resources. A sensor node is comprised of three components: a sensor, a processor and a wireless communication device. These sensor nodes dispersed throughout it to monitor, collect, and transmit data. The sensors are inexpensive, simple, and their power source is irreplaceable. Knowing the sensors power levels cannot be restored, many protocols have been developed to make collecting, receiving and transferring data more energy efficient. In this paper, we propose a multi-hop cluster based routing protocol which is more energy efficient than single hop protocol. Simulation results show that the protocol offers a better performance than single-hop clustering routing protocols in terms of network lifetime and energy consumption by improving FND.

KEYWORDS

Wireless Sensor Network, First Node Death (FND), Multi-Hop Communication, Energy Efficient, Gateway Nodes

1. INTRODUCTION

Wireless sensor networks (WSN's) [1] have gained worldwide attention in recent years, particularly with the proliferation in Micro-Electro-Mechanical Systems (MEMS) technology which has facilitated the development of smart sensors. These sensors are small, with limited processing and computing resources, and they are inexpensive compared to traditional sensors. These sensor nodes can sense, measure, and gather information from the environment and, based on some local decision process, they can transmit the sensed data to the user.

A WSN typically has little or no infrastructure. It consists of a number of sensor nodes (few tens to thousands) working together to monitor a region to obtain data about the environment. These sensors have the ability to communicate either among each other or directly to an external base-station (BS). A greater number of sensors allows for sensing over larger geographical regions with greater accuracy. The sensor sends such collected data, usually via radio transmitter, to a command center (sink) either directly or through a data concentration center (a gateway).

Routing protocol is one of the core technologies in the WSN. Due to its inherent characteristics, routing is full of challenge in WSN [2]. Clustering is a well-know and widely used exploratory data analysis technique, and it is particularly useful for applications that require scalability to hundreds or thousands of nodes [3]. For large-scale networks, node clustering has been proposed for efficient organization of the sensor network topology, and prolonging the network lifetime. Among the sources of energy consumption in a sensor node, wireless data transmission is the most critical. Within a clustering organization, intra-cluster communication can be single hop or multi-hop, as well as inter-cluster communication.

In this paper, we analyze energy efficient multi-hop clustering routing algorithm by a sensor node for WSN. We first describe the new energy based multi-hop with gateway node routing scheme, and then simulation results in MATLAB [4]. Further, the performance analysis of the proposed scheme is compared with benchmark clustering algorithm LEACH [5].

The remainder of this paper is organized as follows: Section 2 describes the related work. Section 3 describes the proposed multi-hop routing scheme. Simulation results are discussed in section 4 and conclusions are drawn in section 5.

2. RELATED WORKS

In sensor networks deployed in harsh or unstructured environments, sensor nodes are typically powered by irreplaceable batteries with a limited amount of energy supply. Ideally we would like the sensor network to perform its functionality as long as possible. Optimal routing maximizes the network functionality by minimizing the total energy consumption and optimizing the network-wide load balance to prolong the lifetime of sensor networks have been an essential task in sensor network implementation.

Routing is a process of determining a path between source and destination upon request of data transmission. A variety of protocols have been proposed to enhance the life of WSN and for routing the correct data to the base station. Employing clustering techniques in routing protocols can hierarchically organize the network topology and prolongs the lifetime of a wireless sensor network, and contributes to overall system scalability. Various protocols [6] like LEACH, HEED, PEGASIS, TEEN, and APTEEN are available to route the data from node to base station in WSN.

A single-hop clustering routing protocol can reduce the communication overhead by selecting a CH to forward data to base station via one hop. Many single-hops clustering routing protocol have been proposed like LEACH and HEED. But when communication distance increases, single hop communication consumes more energy. Multi-hop communication consumes less energy than single hop protocols for long distances. Many multi-hop routing protocols have been proposed like M-LEACH [7] and MR-LEACH [8].

Low Energy Adaptive Clustering Hierarchy (LEACH) is the first clustering protocol that was proposed for reducing power consumption. It forms clusters by using a distributed algorithm, each node has a certain probability of becoming a cluster head per round, and the task of being a cluster head is rotated between nodes. A non-CH node determines its cluster by choosing the CH that can be reached with the least communication energy consumption. In the data transmission stage, each cluster head sends an aggregated packet to the base station by single hop.

LEACH randomly selects a few sensor nodes as CHs and rotates this role to evenly load among the sensors in the network in each round. In LEACH, the cluster head (CH) nodes compress data arriving from nodes that belong to the respective cluster, and send an aggregated packet to the base station. A predetermined fraction of nodes, p, elect themselves as CHs in the following manner. A sensor node chooses a random number, r, between 0 and 1. If this random number is less than a threshold value, T (n), the node becomes a cluster-head for the current round. The threshold value is calculated based on an equation that incorporates the desired percentage to become a cluster-head, the current round, and the set of nodes that have not been selected as a cluster-head in the last (1/p) rounds, denoted by G. It is given by:

$$T(n) = p/1 - p(r \bmod (1/p)) \quad \text{if } n \text{€} G \quad (1)$$

Each elected CH broadcasts an advertisement message to the rest of the nodes in the network that they are the new cluster-heads. A sensor node or non- CH selects the CHs which is nearest to it.

Inter-Intra Cluster Multi hop-LEACH (M-LEACH) [7] is a cluster based routing algorithm. Basic operation of Multi hop-LEACH is similar to LEACH protocol. There are two major modifications in Multi hop-LEACH protocol with respect to LEACH protocol. Multi hoping is applied to both inter cluster and intra cluster communication. Each cluster is composed of one cluster head (CH) and cluster member nodes. The respective CH gets the sensed data from its cluster member nodes, aggregates the sensed information and then sends it to the Base Station through an optimal multi-hop tree formed between cluster heads (CHs) with base station as root node. When the sensor nodes are deployed in regions of dense vegetation or uneven terrain, it may be beneficial to use multi-hop communication among the nodes in the cluster to reach the cluster head. Intra cluster communication performs in the same way like inter cluster communication.

3. THE PROPOSED MULTI-HOP ALGORITHM

Basic operation of proposed multi-hop clustering routing protocol is multi-hop transmission of data from CH to BS. A multilevel hierarchical, data gathering sensor network architecture is used in this scheme. At the lowest level sensor nodes send data to cluster head, and cluster head sends data to gateway nodes. The gateway nodes, which forms the next level of hierarchy, are programmed to communicate with a sink (Base station) located outside from the network field.

3.1. Assumptions

The following **assumptions** are made for the new scheme:

- The Base Station (i.e. data sink) located far away from the sensing field and it is stationary after deployment.
- The Base station (BS) has the information about the location of each node and the location of gateway nodes.
- Nodes are dispersed in a 2-dimensional space and cannot be recharged after deployment.
- Nodes are uniformly distributed in network and they are stationary after deployment.
- Nodes are homogeneous and have the same capabilities. Each node is assigned a unique identifier (ID).
- All nodes can send data to Gateway nodes.
- All nodes have the information of gateway nodes locations via an initial broadcast message.

- 10 Gateway nodes are dispersed in same sensor field at left most upper corner of the network area.
- Gateway nodes have the information of location of BS.
- Gateway Nodes have endless battery power means their batteries can be recharged.
- Gateway nodes have only two responsibilities, one is received data from cluster heads and second is transmit the data to BS.
- A gateway node can connect with only one CH node. The protocol limits that a gateway node can connect with only one CH node.
- Each Sensor node has the same initial power.
- In the first round, each node has a probability p of becoming the cluster head..
- Data compression is done by the Cluster head.
- Energy of transmission depends on the distance (source to destination) and data size.

3.2. Proposed Algorithm

The proposed **algorithm** works in rounds. Each round performs these following steps:

1. Periodically the base station starts a new round by incrementing the round number.
2. Selects cluster heads on the basis of leach protocol with probability 0.1 and the CH should not be more than 10 in number, in each round. In each round a sensor node elects itself as a cluster head by selecting a random number to compare to the threshold value. The threshold $T(n)$ is set as: $T(n) = \{P / 1 - P * (r \bmod 1/P)\}$ if n belongs to G, if not its 0. P is the desired percentage of cluster heads, r is the current round, and G is the set nodes that have not been cluster heads in the last (1/P) rounds.
3. As soon as a CH is formed, it selects a gateway node which lies closest to it.
4. Make Clusters by allocating the cluster head to each node of the network on the basis of minimum distance between nodes to Cluster head (CH).
5. Sensor nodes wake up, senses data, and forwards sensed data to respective CHs.
6. The CHs aggregates data receiving from all cluster members and then send data to the gateway nodes on the basis of one-to-one communication.
7. Now the gateway nodes further send the data to the BS and protocol goes in next round till the last round is not encountered.

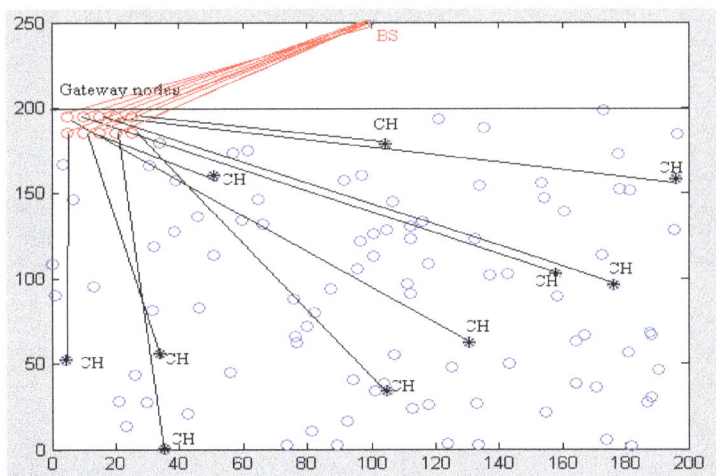

Figure 1. Model view of the proposed scheme

3.3. Flowchart of Proposed Algorithm

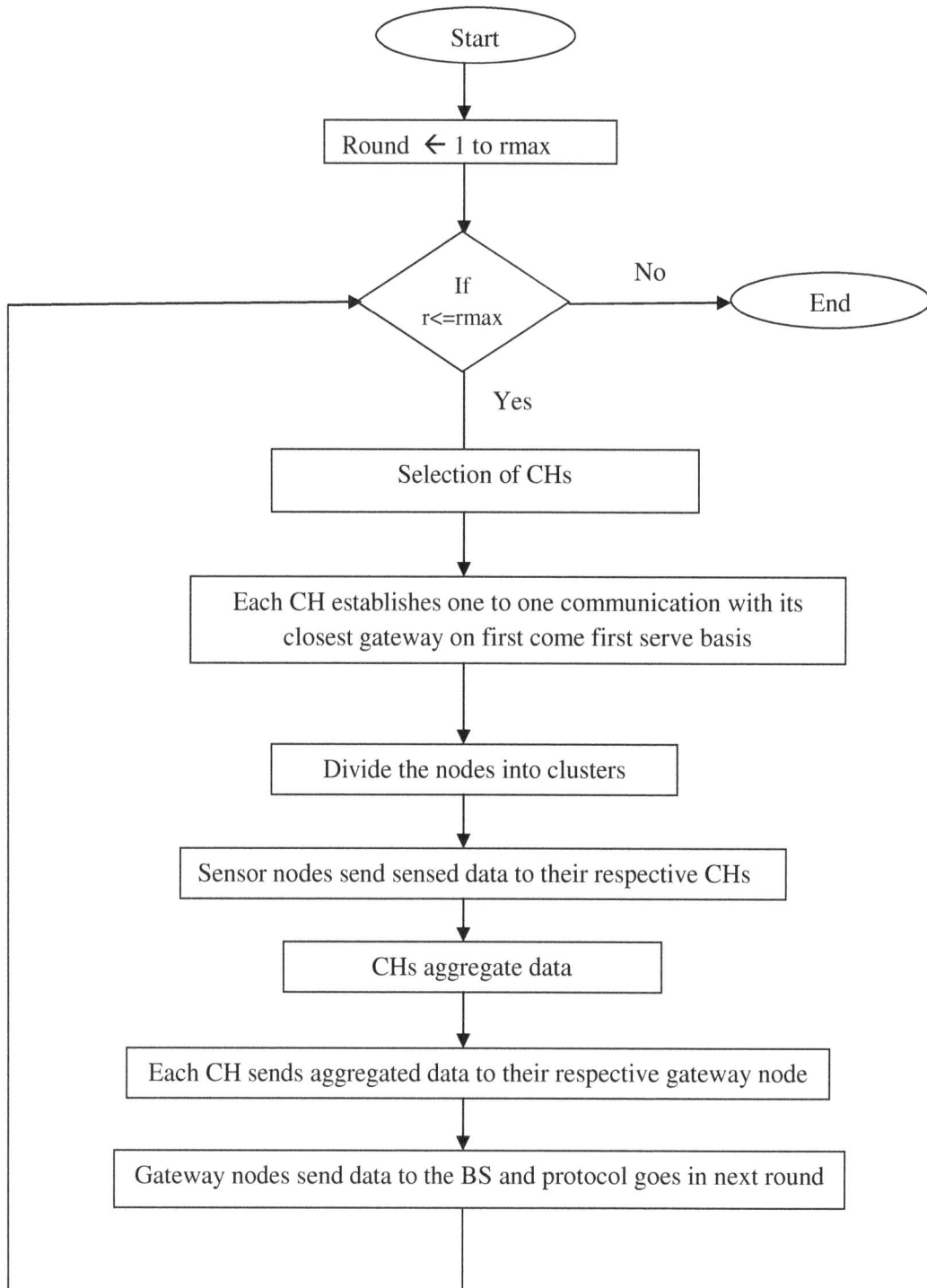

Figure 2. Flow Chart of Proposed Algorithm

3.4. Pseudo Code

- {for each round
- **// Choosing Cluster Head**
- Threshold is set to (P / (1 – P * (round % 1/P)))
- {for each node
- {if number of cluster <= 10 && energy of node > 0
- Assign a random number
- {if (random number < threshold value) && (the node has not been cluster head)
- Node is Cluster head //assign node id to cluster head list
- Increment cluster head count //a new cluster head has been added
- Else go to the next node}
- Else go to the next node}
- }
-
- **// CHs connecting to gateway nodes**
- { for each cluster head
- {for all gateway nodes
- cluster head coordinate x is assigned to x1
- cluster head coordinate y is assigned to y1
- {if gateway node flag is false
- gateway node coordinate x is assigned to x2
- gateway node coordinate y is assigned to y2
- {if it is the first gateway node
- the distance between cluster head and gateway node is the least distance
- gateway node id is assigned as closest gateway node to cluster head
- }
- {else
- Distance between cluster head and current gateway node is current distance
- {if current distance < least distance
- Current distances is now assigned to least distance
- Cluster gateway node id is assigned as closest gateway node to cluster head
- Gateway node flag is set·true
- }
- }
- Else go to next gateway node}
- }
- }
-
- **// Generating Clusters**
- {For each node
- if node is a cluster head
- go to next node

- {else
- {for each cluster head
- node coordinate x is assigned to x1
- node coordinate y is assigned to y1
- cluster head coordinate x is assigned to x2
- cluster head coordinate y is assigned to y2
- {if it is the first cluster head
- the distance between node and cluster head is the least distance
- cluster head id is assigned as closest cluster head to node
- }
- {else
- Distance between node and current cluster head is current distance
- {if current distance < least distance
- Current distances is now assigned to least distance
- Cluster head id is assigned as closest cluster head to node
- }
- }
- }
- }
- }
-
- **//Simulating Transmission and Reception**
- {if distance between node and cluster head is <= the transmission range
- Transmission cost is $E_{Tx}(l, d) = E_{elec} * l + E_{fs} * l * d^2$
- Reception cost is $E_{Rx}(l) = Eelec * l$
- Subtract the transmission cost from the sending node
- {if remaining energy <= 0
- display node has died
- exit the program
- }
- Subtract the reception cost from the receiving node
- {if remaining energy <= 0
- display node has died
- exit the program
- }
- return the sum of transmission cost and reception cost and calculate the residual energy of each node
- }
- } //rmax

4. SIMULATION AND PERFORMANCE EVALUATIONS

All simulations have been implemented using MATLAB. Assuming that 100 nodes are randomly distributed in field of 200x200 and the sink is located about 50m away from the field edge. The simulation parameters are given in Table 1. The performance of the proposed multi-hop protocol scheme is compared with that of the single hop Leach protocol.

4.1. Energy Model for Communication

We assume a simple model for the radio hardware energy dissipation where the transmitter dissipates energy to run the radio electronics and the power amplifier, and the receiver dissipates energy to run the radio electronics. For the experiments described here only the free space channel model is used. Thus, to transmit an l-bit message a distance d, the radio expends energy:

$$E_{Tx}(l, d) = (l\, E_{elec} + l\, E_{fs}\, d^2)\qquad(2)$$

To receive this message, the radio expends energy:

$$E_{Rx}(l) = l\, E_{elec}\qquad(3)$$

4.2. Simulation Parameter

Table 1. Simulation Parameters

Parameter	Values
Simulation Round	2000
Sink Location	(100,250)
Network Size	200 x 200
Number of nodes	100
Number of Gateway nodes	10
CH probability	0.1
Fusion rate (cc)	0.6
Initial node power	0.5 Joule
Nodes Distribution	Nodes are uniformly distributed
Control Packet Size	500 bits
Data Packet size	4000 bits
Energy dissipation (Efs)	10*0.000000000001 Joule
Energy for Transmission (E_{TX})	50*0.000000000001 Joule
Energy for Reception (E_{RX})	50*0.000000000001 Joule
Energy for Data Aggregation (EDA)	5*0.000000000001 Joule

4.3. Simulation Results

4.3.1. Network Life Time

When a node is dead in the network it'll not be the part of the network. It shows that if a dead node occurs in early rounds of the algorithm, this may affect lifespan of the network or drag towards the early dead of all nodes. Table 2 shows the simulation results of the two schemes. Fig.3 concludes that in the proposed algorithm, the first node dies later in the network.

Table 2. Network Life Time (First node dead)

No. of Simulation Runs	Round Number when first node dies	
	Leach	Proposed scheme
1	302	365
2	312	395
3	298	343
4	282	384
5	309	343
6	272	382
7	262	329
8	294	348
9	261	380
10	295	362

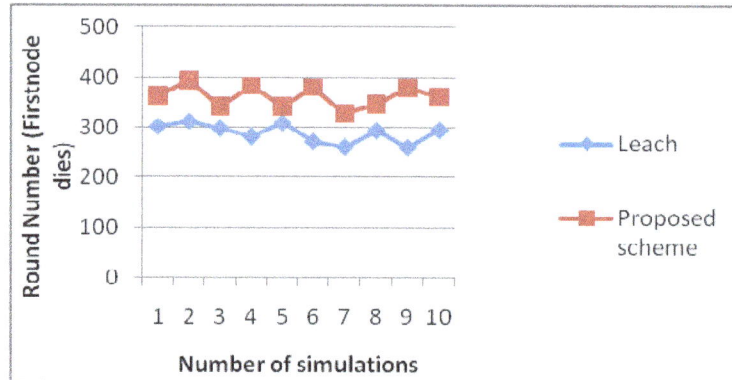

Figure 3. Network Life Time (First node dead) v/s No. of Simulation run

4.2.2. Network Lifetime with Number of Alive Nodes

More alive nodes contribute to the increase in network life time. Table3 and Figure4 show the number of nodes alive in the network with the increase in number of rounds. It is vivid that the lifetime of WSN using multi-hop proposed scheme is better compared to Leach Protocol.

Table 3. Network Life Time with Number of Alive nodes

Round Number	Number of Alive Nodes	
	Leach	Proposed scheme
100	100	100
300	92	100
600	59	76
900	27	46

1200	24	38
1500	15	30
1800	13	21
2100	5	17

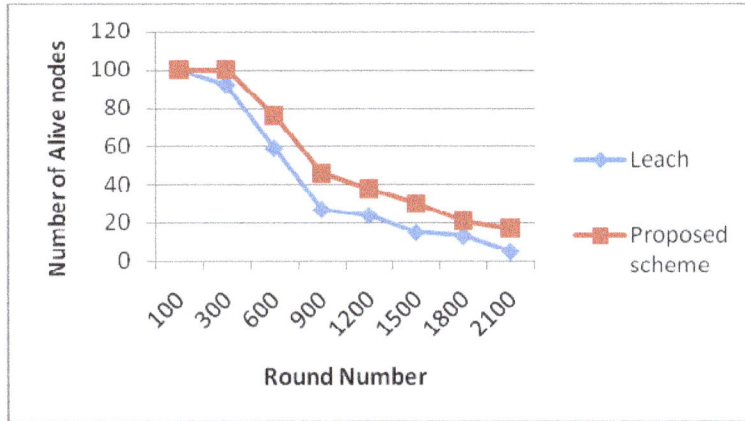

Figure 4. Network Life Time (Alive nodes) v/s Number of Rounds

4.2.3. Network Life Time with Varying Packet Size

Even on varying the packet size the network lifetime for the proposed algorithm remains better than that of the Leach. Table 4 and Fig.5 show the results on comparison.

Table 4. Network Life Time with different Packet Size

Packet Size	Round Number when first node dies	
	leach	Proposed Scheme
10000	111	136
9000	133	148
8000	134	169
7000	156	220
6000	193	274
5000	216	342
4000	295	380
3000	463	633
2000	615	744
1000	1179	1265

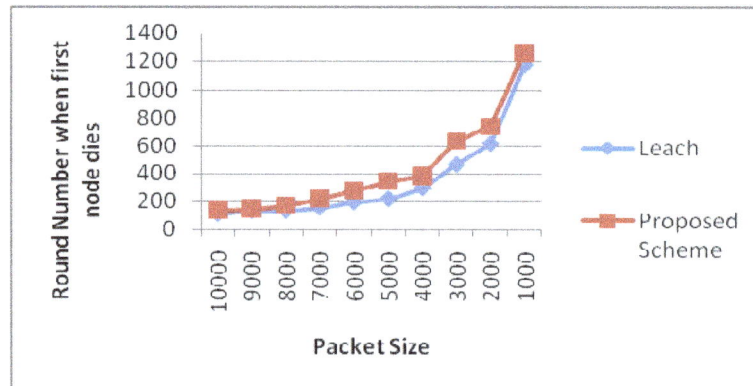

Figure 5. Network Life Time with varying packet size

5. CONCLUSIONS

Energy consumption is the main design issue in routing of wireless Sensor Network. We concluded that energy consumed for single hop transmission is more than multi-hop transmission for long distances. A new multi-hop routing protocol for the homogeneous wireless sensor networks has been presented and the performance of the system is evaluated to minimize the energy consumption and increase the life time of sensor network. The simulation results reveal that the LEACH protocol consumes more energy and the network has shorter lifetime than proposed multi-hop protocol with gateway nodes. We have determined the impact of packet length on the network lifetime. Finally, simulation results indicate that proposed protocol can more efficiently balance energy consumption of an entire network and thus extends the network lifetime. The proposed protocol is for the homogeneous network and we propose to extend this work for heterogeneous network in future work.

REFERENCES

[1]   Wireless sensor network survey Jennifer Yick, Biswanath Mukherjee, Dipak Ghosal, Department of Computer Science, University of California, Davis, CA 95616, United States

[2]   H. W. Kim, H. S. Seo(2010), "Modeling of Energy-efficient Applicable Routing Algorithm in WSN", International Journal of Digital Content Technology and its Applications, vol. 4, no. 5, pp.13-22.

[3]   W. B. Hcinzclman, A. P. Cnandrakasan,(2002) "An application-specific protocol architecture for wireless microsensor networks," IEEE Transactions on Wireless Communications, vol. 1, no. 4, pp.660–670.

[4]   http://www.mathworks.in

[5]   Wendi Rabiner Heinzelman (2000)."Energy-Efficient Communication Protocol for Wireless Microsensor Networks". In Proceeding of the 33rd Hawaii International Conference on System Sciences,pp1-10

[6]   Ankita Joshi , Lakshmi Priya.M "A Survey of Hierarchical Routing   Protocols in Wireless Sensor Network", International Conference on Information Systems

[7]   Rajashree.V.Biradar , Dr. S. R. Sawant , Dr. R. R. Mudholkar , Dr. V.C .Patil (2011)" Inter-Intra Cluster Multihop-LEACH Routing In Self-Organizing Wireless Sensor Networks", International Journal of Research and Reviews in Computer Science (IJRRCS) Vol. 2, No. 1.

[8]   Muhamnmad Omer Farooq, Abdul Basit Dogar, Ghalib Asadullah Shah (2010)" MR-LEACH: Multi-hop Routing with Low Energy Adaptive Clustering Hierarchy", Fourth International Conference on Sensor Technologies and Applications.

[9]   Guihai Chen · Chengfa Li · Mao Ye · JieWu," An unequal cluster-based routing protocol in wireless   sensor networks", Wireless Netw DOI 10.1007/s11276-007-0035-8

[10]   Seema Bandyopadhyay and Edward J. Coyle," An Energy Efficient Hierarchical Clustering Algorithm for Wireless Sensor Networks".

[11]   F. Akyildiz, W. Su, (2002) "A survey on sensor networks", Communications Magazine, IEEE, vol. 40, no. 8, pp.102-114.

6

INTEGRATED SOLUTION SCHEME FOR HANDOVER LATENCY DIMINUTION IN PROXY MOBILE IPv6

Md. Mahedi Hassan and Poo Kuan Hoong

Faculty of Information Technology, Multimedia University, Cyberjaya, Malaysia
mahedi822002@gmail.com, khpoo@mmu.edu.my

ABSTRACT

Recent trends show that there are swift developments and fast convergence of wireless and mobile communication networks with internet services to provide the quality of ubiquitous access to network users. Most of the wireless networks and mobile cellular networks are moving to be all IP based. These networks are connected through the private IP core networks using the TCP/IP protocol or through the Internet. As such, there is room to improve the mobility support through the Internet and support ubiquitous network access by providing seamless handover. This is especially true with the invention of portable mobile and laptop devices that can be connected almost everywhere at any time. However, the recent explosion on the usage of mobile and laptop devices has also generated several issues in terms of performance and quality of service. Nowadays, mobile users demand high quality performance, best quality of services and seamless connections that support real-time application such as audio and video streaming. The goal of this paper is to study the impact and evaluate the mobility management protocols under micro mobility domain on link layer and network layer handover performance. Therefore, this paper proposes an integration solution of network-based mobility management framework, based on Proxy Mobile IPv6, to alleviate handover latency, packet loss and increase throughput and the performance of video transmission when mobile host moves to new network during handover on high speed mobility. Simulations are conducted to analyze the relationship between the network performances with the moving speed of mobile host over mobility protocols. Based on simulation results, we presented and analyzed the results of mobility protocols under intra-domain traffics in micro mobility domain.

KEYWORDS

Seamless Handover, Mobility Protocols, Proxy Mobile IPv6, PMIPv6, Video Transmission, NS-2

1. INTRODUCTION

In recent years, mobile and wireless communications have undergone tremendous changes due to the rapid development in wireless and communication technologies as well as the ever increasing demands by users. Nowadays, mobile end-users are constantly on the go and most of the time, they are moving from one place to another place in rapid pace. As a result, connected mobile devices are also constantly changing their points of attachment to the communication networks, such as Mobile Cellular Networks (MCN), Wireless Local Area Networks (WLAN), Wireless Personal Access Networks (WPAN), and so on. These days, most of the wireless and mobile communication networks are moving towards all IP based. These communication networks are either connected together through the Internet or through private IP core networks. In order to maintain connection, one of the main challenges faced by Mobile Host (MH) is the ability to obtain a new IP address and update its communication partners, while moving amongst these different wireless and mobile networks.

In order to meet the above challenge, Internet Engineering Task Force (IETF) [1] designed a new standard solution for Internet mobility officially called – IPv6 mobility support and popularly named as Mobile IPv6 (MIPv6) [2]. MIPv6 is the modified version of MIPv4, that has great practicality and able to provide seamless connectivity to allow a mobile device to maintain established communication sessions whilst roaming in different parts of the Internet.

When a MH is handed over from one network to another network, it changes the point of attachment from one access router (AR) to another. This is commonly known as *handover* which allows MH to establish a new connection with a new subnet. Handover is also defined as the process of changing between two ARs and when ARs' point of attachment in the network changes. The point of attachment is a base station (BS) for cellular network, or an AR for WLAN. Commonly, handover can be handled in the link layer, if both the ARs are involved in the same network domain. Otherwise, a route change in the IP layer possibly will be needed the so-called network layer handover. In this case, Mobile IPv6 is a standard protocol for handling network layer handover.

For IP-mobility protocols, the IP handover performance is one of the most important issues that need to be addressed. IP handover occurs when a MH changes its network point of attachment from one BS to another. Some of the major problems that may occur during handover are handover latency and packet loss which can degrade the performance and reduce quality of service. In a nutshell, handover latency is the time interval between the last data segment received through the previous access point (AP) and first data segment received through the next AP [3]. The major problem arises with handovers is the blackout period when a MH is not able to receive packets, which causes a high number of packet loss and communication disruption. Such long handover latency might disrupt ongoing communication session and some interruptions. If that change is not performed efficiently, end-to-end transmission delay, jitters and packet loss will occur and this will directly impact and disrupt applications perceived quality of services. For example, handovers that might reach hundreds of milliseconds would not be acceptable for delay-sensitive applications like video streaming and network gaming [3] [4].

Currently, there are several mobility protocols which have been proposed in order to alleviate such performance limitations. One of which is the enhanced version of terminal independent Mobile IP (eTIMIP) [5] [6], which is a kind of mobility management protocol. eTIMIP enhances the terminal independent Mobile IP (TIMIP) [5] by reducing the amount of latency in IP layer mobility management messages exchanged between an MH and its peer entities, and the amount of signaling over the global Internet when a MH traverses within a defined local domain. TIMIP is an example of IP based micro-mobility protocol that allows MH with legacy IP stacks to roam within an IP domain and doesn't require changes to the IP protocol stack of MH in a micro mobility domain.

Compared to the above mobility protocols, Proxy Mobile IPv6 (PMIPv6) [7] defines a domain in which the MH can roam without being aware of any layer 3 (L_3) movement since it will always receive the same network prefix in the Router Advertisement (RA). The PMIPv6 specification defines a protocol to support Network-based Localized Mobility Management (NETLMM) [7] [8] where the MH is not involved in the signaling. This new approach is motivated by the cost to modify the protocol stack of all devices to support Mobile IP and potentially its extensions and to support handover mechanisms similar to the ones used in 3GPP/3GPP2 cellular networks.

We make use of Network Simulator, ns-2 [9] in this paper to simulate, examine and compare the performances of eTIMIP, TIMIP, PMIPv6 as well as our proposed integrated solution of PMIPv6 with MIH and Neighbor Discovery (PMIPv6-MIH) in intra-domain traffic with high speed MH. We evaluate the handover latency, packet loss, Peak Signal-to-Noise Ratio (PSNR) [10] and packet delivery throughput of video transmission over PMIPv6 and our proposed integrated solution of PMIPv6-MIH, and also compare the handover latency and packet delivery throughput of transmission control protocol (TCP) and user datagram protocol (UDP) for eTIMIP, TIMIP, PMIPv6 and PMIPv6-MIH in intra-domain traffic.

The rest of this paper is structured as follows: Section 2 briefly explain related research works on the mobility protocols. Section 3 explains overview of media independent handover. Section 4 briefly describes the propose solution scheme. Section 5 shows simulation results of UDP and TCP flow under intra-domain traffic and video transmission over PMIPv6. Finally, Section 6 we conclude the paper and provide possible future works.

2. RESEARCH BACKGROUND

For mobility protocols, there are several protocols to reduce handover latency and packet loss, such as the Session Initiation Protocol (SIP) [11] and the Stream Control Transmission Protocol (SCTP) [12]. Both protocols focus on mobility management on an end-to-end basis but they don't have the potential to achieve short handover latency in network layer. The communication sessions in these protocols are initiated and maintained through servers. The behaviour of these protocols is similar to the standard Mobile IP scheme during handovers. However, there are some enhanced Mobile IP schemes that able to reduce the handover latency such as PMIPv6 and CIMS, (Columbia IP Micro-Mobility Suite) [13].

2.1. Micro Mobility Protocols

Micro mobility protocols work within an administrative domain which is to ensure that packets are arriving from the internet and addressed to the MHs that forward to the appropriate wireless AP in an efficient manner. It is also called intra-domain traffic [14]. Under the CIMS (Columbia IP Micro-Mobility Suite) project, several micro mobility protocols have been proposed such as –Handoff-Aware Wireless Access Internet Infrastructure (Hawaii) and Cellular IP (CIP).

The CIMS is an extension that offers micro-mobility support. CIMS implements HMIP (Hierarchical Mobile IP) and two micro-mobility protocols for CIP and Hawaii. The CIMS project is mainly focused on intra-domain handover and uses the basic idea of Mobile IP for inter-domain handover.

Subsequently, the CIMS project was enhanced by Pedro et. al. [15] which included the original implementation of TIMIP protocol, and the extended version of TIMIP protocol such as eTIMIP as well as the implementation of CIP, HAWAII, and HMIP protocols. The proposed eTIMIP protocol which is a mobility solution protocol that provides both network and terminal independent mobile architectures based on the usage of overlay micro-mobility architecture.

2.2. Enhanced version of Terminal Independent Mobile IP (eTIMIP)

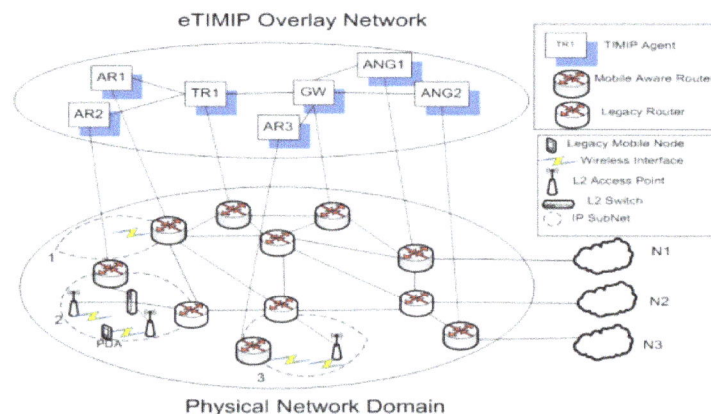

Figure 1. Architecture of eTIMIP

The physical network and overlay network are two complementary networks that are organized in the architecture of eTIMIP. Both networks are separated in the mobile routing from the traditional intra-domain routing which also known as fixed routing. Generally, the physical network can have any possible topology, where it is managed by any specialized fixed routing protocol. The overlay network is used to perform the mobile routing, where it selects routers which support the eTIMIP agents, in which will be organized in a logical tree that supports multiple points of attachment to the external of the domain.

2.3. Proxy Mobile IPv6 (PMIPv6)

PMIPv6 is designed to provide an effective network-based mobility management protocol for next generation wireless networks that main provides support to a MH in a topologically localized domain. In general terms, PMIPv6 extends MIPv6 signaling messages and reuse the functionality of HA to support mobility for MH without host involvement. In the network, mobility entities are introduced to track the movement of MH, initiate mobility signaling on behalf of MH and setup the routing state required. The core functional entities in PMIPv6 are the Mobile Access Gateway (MAG) and Local Mobility Anchor (LMA). Typically, MAG runs on the AR. The main role of the MAG is to perform the detection of the MH's movements and initiate mobility-related signaling with the MH's LMA on behalf of the MH. In addition, the MAG establishes a tunnel with the LMA for forwarding the data packets destined to MH and emulates the MH's home network on the access network for each MH. On the other hand, LMA is similar to the HA in MIPv6 but it is the HA of a MH in a PMIPv6 domain. The main role of the LMA is to manage the location of a MH while it moves around within a PMIPv6 domain, and it also includes a binding cache entry for each currently registered MH and also allocates a Home Network Prefix (HNP) to a MH. An overview of PMIPv6 architecture is shown in figure 2.

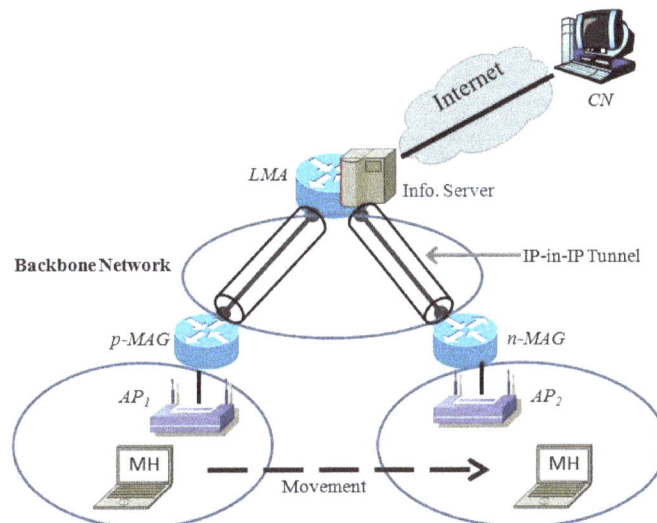

Figure 2. Architecture of PMIPv6

Since the PMIPv6 was only designed to provide local mobility management, it still suffers from a lengthy handover latency and packet loss during the handover process when MH moves to a new network or different technology with a very high speed. Even more, since detecting MHs' detachment and attachment events remains difficult in many wireless networks, increase handover latency and in-fly packets will certainly be dropped at new MAG (n-MAG).

3. OVERVIEW OF MEDIA INDEPENDENT HANDOVER

The working group of IEEE 802.21 [15] developed a standard specification, called Media Independent Handover (MIH), which defines extensible media access independent mechanisms that facilitate handover optimization between heterogeneous IEEE 802 systems such as handover of IP sessions from one layer 2 (L_2) access technologies to another. The MIH services introduce various signaling, particularly for handover initiation and preparation and to help enhance the handover performance.

The MIH services introduce various signaling, particularly for handover initiation and preparation and to help enhance the handover performance. Figure 3 shows the overall framework of MIH.

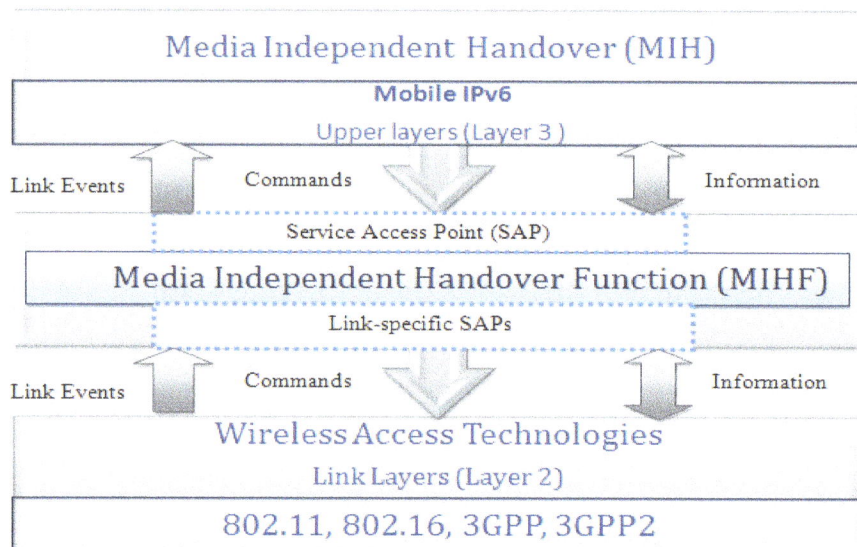

Figure 3. The framework of MIH services

Basically, IEEE 802.21 introduces three different types of communications with different associated semantics [16], the so-called MIH services: Media Independent Event Service (MIES), Media Independent Command Service (MICS) and Media Independent Information Service (MIIS).

MIES introduces event services that provide event classification, event filtering and event reporting corresponding to dynamic changes in link characteristics, links status, and link quality. It also helps to notify the MIH users (MIHU) such as PMIPv6 about events happening at the lower layers like link down, link up, link going down, link parameters report and link detected etc and essentially work as L_2 triggers.

MICS provides the command services that enable the MIH users to manage and control link behavior relevant to handovers and mobility, such as force change or handover of an interface. The commands generally carry the upper layers like L_3 decisions to the lower layers like L_2 on the local device entity or at the remote entity. There are several examples of MICS commands, such as MIH scan, MIH configure, MIH handover initiate, MIH Handover prepare and MIH handover complete.

MIH provides information about the characteristics and services through a MIIS which enables effective handover decisions and system access based on the information about all networks from any single L_2 networks. MIIS provides registered MIH users with the knowledge base of the network and information elements and corresponding query-response mechanisms for the transfer of information. By utilizing these services, the MIH users are able to enhance handover performance such as through informed early decisions and signaling. MIIS are classified into three groups, namely general or access network specific information, Point of Attachment (PoA) specific information and vendor specific information.

4. PROPOSED INTEGRATED SOLUTION

In response to the PMIPv6 problems mentioned in Section 2, we proposed solution scheme that provides an integrated solution with integrate the analysis of handover latency introduced by PMIPv6 with the seamless handover solution used by MIH as well as the Neighbor Discovery message of IPv6 to reduce handover latency and packet loss on network layer at n-MAG to avoid the on-the-fly packet loss during the handover process. Figure 4 represents the proposed integrated solution of PMIPv6-MIH.

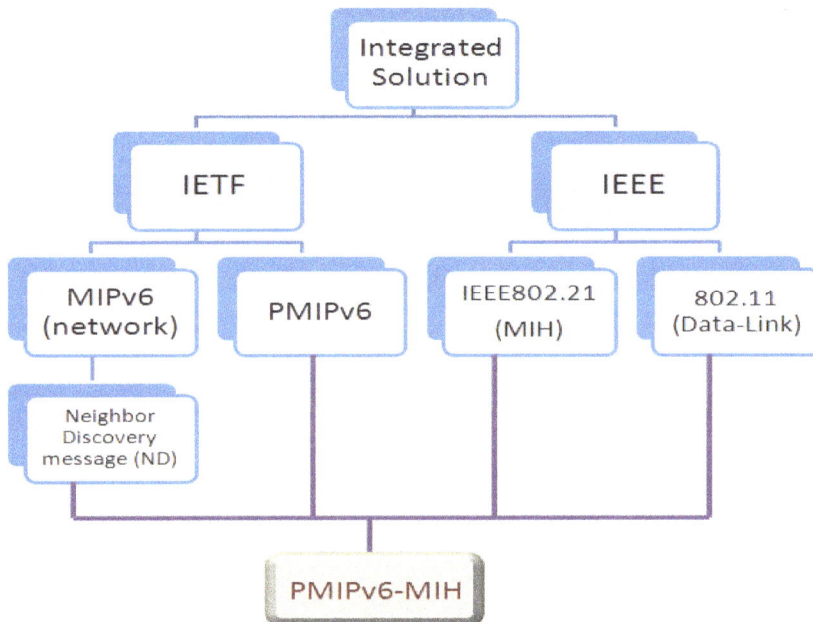

Figure 4. Proposed Integrated Solution

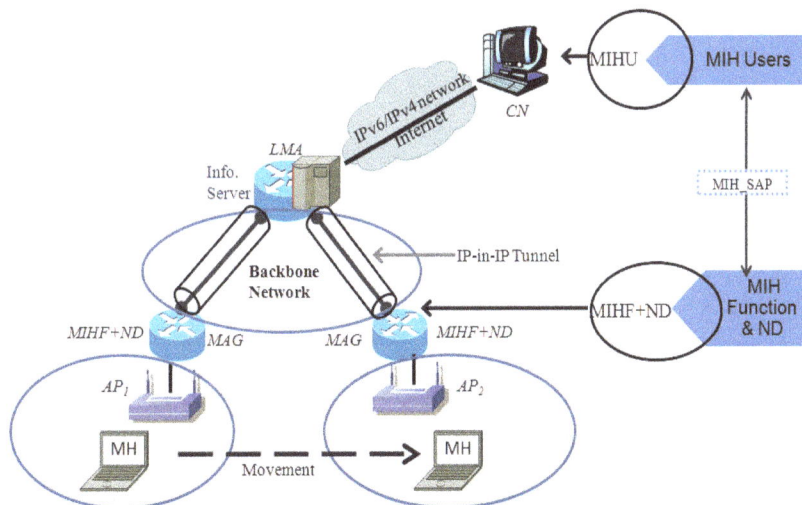

Figure 5. Integrated solution architecture of PMIPv6

Figure 5 presents the key functionality is provided by Media Independent Handover (MIH) which is communication among the various wireless layers and the IP layer. The working group of IEEE 802.21 introduces a Media Independent Handover Function (MIHF) that is situated in the protocol stack between the wireless access technologies at lower layer and IP at upper layer. It also provides the services to the L_3 and L_2 through well defined Service Access Points (SAPs) [16].

4.1. Neighbor Discovery

Neighbor Discovery (ND) enables the network discovery and selection process by sending network information to the neighbor MAG before handover that can helps to eliminate the need for MAG to acquire the MH-profile from the policy server/AAA whenever a MH performs handover between two networks in micro mobility domain. It avoids the packet loss of on-the-fly packet which is routed between the LMA and previous MAG (p-MAG). This network information could include information about router discovery, parameter discovery, MH-profile which contains the MH-Identifier, MH home network prefix, LMA address (LMAA), MIH handover messages etc., of nearby network links.

4.2. Analysis of Handover Latency and Assumptions

The overall handover latency consists of the L_2 and L_3 operations. The handover latency is consequent on the processing time involved in each step of handover procedure on each layer.

The handover latency ($L_{seamless}$) can be expressed as:

$$Lseamless = L_{L_3} + L_{L_2} \quad\quad .... (1)$$

where L_{L3} represents the network layer as example switching latency and L_{L2} represents link layer as example switching time.

On L_3, the handover latency is affected by IP connectivity latency. The IP connectivity latency results from the time for movement detection (MD), configure a new CoA (care-of-address), Duplicate Address Detection (DAD) and binding registration. Therefore, L_3 can be denoted as follows:

$$L_3 = T_{config} + T_{DAD} + T_{reg} + T_{move} \quad\quad .... (2)$$

where T_{move} represents the time required for the MH to receive beacons from n-MAG, after disconnecting from the p-MAG. In order to estimate the movement detection delay, based on the assumptions of mobility management protocols that the times taken for MD are RS and RA messages as follow:

$$T_{move} = T_{RS} + T_{RA} \quad\quad .... (3)$$

T_{conf} represents the time that taken for new CoA configuration. T_{reg} represents the time elapsed between the sending of Proxy Binding Update (PBU) from the MAG to the LMA and Proxy Binding Advertisement (PBA) from the LMA to the MAG and the arrival/transmission of the first packet through the n-MAG. Binding registration is the sum of the round trip time between MAG and LMA and the processing time as follows:

$$T_{reg} = T_{PBU} + T_{PBA} \quad\quad .... (4)$$

T_{DAD} represents the time required to recognize the uniqueness of an IPv6 address. Once the MH discovers a new router and creates a new CoA it tries to find out if the particular address is unique. This process is called DAD and it is a significant part of the whole IPv6 process.

As simplification of (2), (3) and (4) equations, it can be expressed as:

$$L_3 = T_{config} + T_{DAD} + T_{PBU} + T_{PBA} + T_{RS} + T_{RA} \quad\quad ....(5)$$

On L_2, MH has to perform three operations during the IEEE 802.11 handover procedure such as scanning (T_{scan}), authentication (T_{AAA}) and re-association (T_{re-ass}). Handover latency at L_2 can be denoted as follows:

$$L_2 = T_{scan} + T_{AAA} + T_{re-ass}$$ ....(6)

T_{scan} represents the time that taken the MH performs a channel scanning to find the potential APs to associate with. When MH detects link deterioration, it starts scanning on each channel finding the best channel based on the Received Signal Strength Indicator (RSSI) value.

T_{AAA} represents the time taken for authentication procedure that depends on the type of authentication in use. The authentication time is round trip time between MH and AP.

While T_{re-ass} represents the time needed for re-association consists of re-association request and reply message exchange between MH and AP if authentication operation is successful.

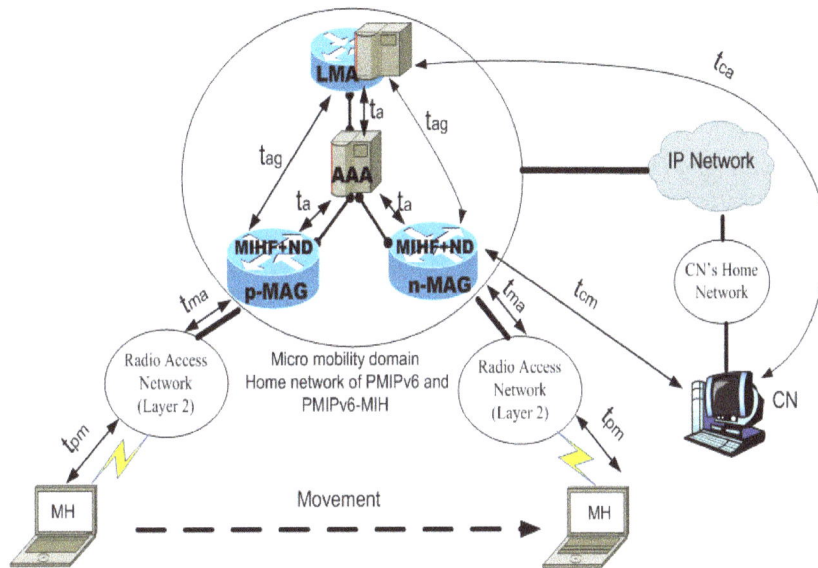

Figure 6. An Analytical Model of PMIPv6 & Integrated solution of PMIPv6-MIH

The following notations are depicted in figure 6 for PMIPv6 & integrated solution of PMIPv6-MIH.

- The delay between the MH and AP is t_{pm}, which is required the time for a packet send between the MH and AP through a wireless link.
- The delay between the AP and n-MAG is t_{ma}, which is the time between the AP and the n-MAG connected to the AP.
- The delay between the n-MAG and LMA is t_{ag}.
- The delay between the LMA and Corresponding Node (CN) is t_{ca}.
- The delay between the n-MAG and CN is t_{cm}, which is the time required for a packet to be sent between the n-MAG and the CN.
- The delay between the mobility agents and AAA is t_a.

As shown in figure 6, we propose an integrated solution of PMIPv6 with MIH and ND to reduce handover latency as the time taken for scanning by informing the MH about the channel information of next APs and use ND message of IPv6 to reduce handover delay and packet loss on network layer at n-MAG to avoid the on-the-fly packet loss during the handover process. Therefore, the handover delay of PMIPv6 and PMIPv6-MIH in a micro mobility domain as follows:

Registration delay:

$$T_{PBU} = t_{ag}$$
$$T_{PBA} = t_{ag}$$

We can add the above equations,

$$T_{PBU} + T_{PBA} = 2t_{ag}$$

Movement Detection delay:

$$T_{RA} = t_{pm} + t_{ma}$$

Authentication delay:

$$T_{AAA} = 2(query + reply) = 2(t_a + t_a) = 4t_a$$

Attachment notification delay:

The attachment notification delay due to the packet from the AP that informs the MAG of an MH's attachment denote T_{attach}

Configuration and DAD delay:

PMIPv6 does not require T_{config} and T_{DAD} because MH is already in the PMIPv6 domain. Once the MH has entered and is roaming inside the PMIPv6 domain, T_{config} is not relevant since according to the PMIPv6 specification, the MH continues to use the same address configuration. A T_{DAD} is required for a link-local address since address collision is possible between MH, MAG and all MHs attached to the same MAG. The T_{DAD} may significantly increase handover delay and is a very time consuming procedure. Typically, T_{DAD} is around one second and sometimes can be much more than that. Therefore, PMIPv6 introduces a per-MH prefix model in which every MH is assigned a unique HNP. This approach may guarantee address uniqueness. The new IP address configuration and the DAD operation for global address are appreciable only when the MH first enters the PMIPv6 domain or move to new PMIPv6 domain.

On L_3 and L_2 equations, Handover delay in PMIPv6 in a micro mobility domain can be expressed as:

$$L_{3_{PMIPv6}} = T_{attach} + 2t_{ag} + t_{pm} + t_{ma} \quad \dots\dots\dots (7)$$

$$L_{2_{PMIPv6}} = T_{scan} + 4t_a + T_{re-ass} \quad \dots\dots\dots (8)$$

During the IEEE802.11 handover procedure the MH performs scanning on the certain number of channels to find the potential APs to associate with. By informing the MH about the channel information of next APs can significantly reduce the scanning time.

However, the scanning time also depends on the type of scanning is used. There are two types of scanning which are defined as active and passive. In active scan mode, MH sends probe request and receives probe response if any AP is available on certain channel. While in passive scan mode, each MHs listens the channel for possible beacon messages which are periodically generated by APs. The handover delay in active scan mode is usually less than in passive scan mode. The operation of passive scan mode depends on the period of beacon generation interval. Therefore, this can provide better battery saving than active scan mode of operation.

As in L_2 trigger, the p-MAG has already authenticated the MH and sends the MH's profile which contains MH-Identifier to the n-MAG through the ND message since the MH is already in the PMIPv6 domain and receiving as well as sending information to CN before the handover. Hence, the authentication delay is eliminated during actual handover. Thus, the L_2 handover delay can be expressed as:

$$L_{2_{PMIPv6-MIH}} = T_{scan} + T_{re-ass} \qquad \ldots\ldots\ldots\ldots\ldots (9)$$

As the parts of L_3 handover delay that should be taken into consideration in PMIPv6. Since we proposed the integrated solution of PMIPv6 with MIH services and ND, the number of handover operations should not be considered for overall handover latency. As a result, L_3 handover delay is considered only two things in integrated solution of PMIPv6-MIH in a micro mobility domain.

  o When MH attaches to the n-MAG and delivers event notification of MIH_Link_up indication, n-MAG sends a PBU message to the LMA for updating the lifetime entry in the binding cache table of the LMA and triggering transmission of buffer data for the MH
  o RA message

Therefore, the overall handover delay at L_3 can be expressed as:

$$L_{3_{PMIPv6-MIH}} = T_{PBU} + T_{RA} \qquad \ldots\ldots\ldots\ldots (10)$$

Based on Analytical model the equations (9) and (10) can be expressed as:

$$L_{2_{PMIPv6-MIH}} = T_{scan} + T_{re-ass}$$

$$L_{3_{PMIPv6-MIH}} = T_{PBU} + T_{RA} = t_{ag} + t_{pm} + t_{ma}$$

Seamless Handover Latency of PMIPv6 and integrated solution of PMIPv6 with MIH+ND can be expressed as:

$$L_{seamless_{(PMIPv6)}} = L_{L_{2PMIPv6}} + L_{L_{3PMIPv6}}$$

$$L_{seamless_{(PMIPv6)}} = T_{attach} + 2t_{ag} + t_{pm} + t_{ma} + T_{scan} + 4t_a + T_{re-ass}$$

$$L_{seamless_{(PMIPv6\ MIH)}} = L_{L_{2PMIPv6-MIH}} + L_{L_{3PMIPv6-MIH}}$$

$$L_{seamless_{(PMIPv6-MIH)}} = t_{ag} + t_{pm} + t_{ma} + T_{scan} + T_{re-ass}$$

4.3. Integrated Solution Scheme

The handover latency is mainly caused by the authentication procedure, attachment notification, obtain MH profile and reconfiguration of the default router when the MH access to a new access network. Therefore, we use MIH and Neighbor Discovery message of IPv6 in order to reduce the handover latency, packet loss and increase throughput.

Figure 7 depicts the control and signaling data flow of the proposed integrated solution scheme of PMIPv6-MIH.

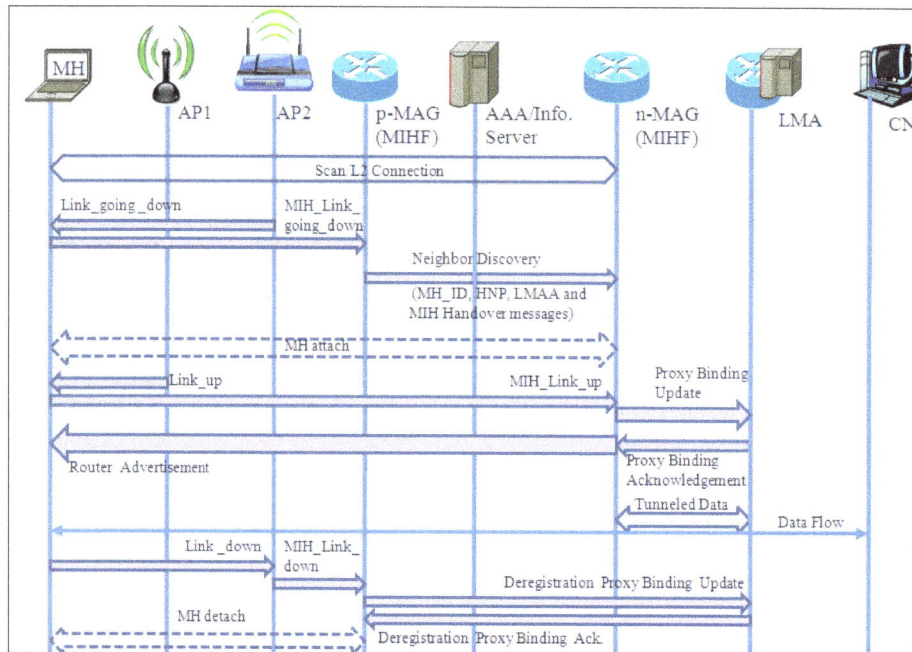

Figure 7. Control and Signaling Data Flow of PMIPv6-MIH

As shown in figure 7, when p-MAG receives a MIH_LGD (Link_going_down) trigger, it transmits the MH profile to its n-MAG using the Neighbor Discovery message of IPv6. This Neighbor Discovery message contains the MH profile, including the MH HNP, MH-ID, and LMAA (LMA address) and MIH handover messages. This eliminates the need for the MAG to acquire the MH profile from the policy server/AAA server whenever an MH performs a handover. The p-MAG starts buffering the packets for the MH in order to avoid the on-the-fly packet loss.

During MH's attachment to the n-MAG, A MIH_Link_up event will be delivered triggering PBU transmission to LMA. After reception of PBU, LMA updates the lifetime for the MH's entry in binding cache table and starts transmission of the buffered data through the tunnel between LMA and MAG.

5. SIMULATION EXPERIMENTS AND RESULTS

In order to examine, evaluate and compare the impact on micro mobility domain handover performance of the proposed integrated solution, simulations were performed to compare and evaluate micro mobility protocols by using the ns-2 [9]. The simulation results are conducted in two ways:

1. For the first simulation, two important performance indicators are measured which are the throughput for packet delivery and handover latency. In order to obtain reasonable results, we measure the performance for micro mobility protocol in intra-domain traffic for both TCP and UDP packet flow.

2. While the second simulation results are conducted in video transmission under four important performance metrics that are the Handover Latency, Throughput, Packet Loss and the quality of video streamed which is measured by PSNR. For the PSNR measurements, tools provided by Chih-Heng Ke [17] are used to simulate video

transmission simulation over wireless network in ns-2. For our simulations, we utilized the tools of video transmission and used a video stream of MPEG-coded with MPEG4 type frames that is used as a source model for MPEG4 traffic. There are two format sizes such as QCIF (176 x 144) and CIF (352 x288), where video frame size is the only difference. For our simulations, we convert a video clip file named "Transformers 2" in CIF format and measure the performance for network-based mobility management protocols in a micro mobility domain with high speed movement of the MH.

5.1. Simulation Scenario Setup

The simulation scenario setup was implemented as a network-based mobility management solution in the simulation of mobility across overlapping wireless access networks in micro mobility domain. The proposed integrated solution scenario setup is the same as the PMIPv6 but further incorporates MIH functionality in the MH and the MAGs. Thus, the simulation setup scenario is as shown in figure 8 below:

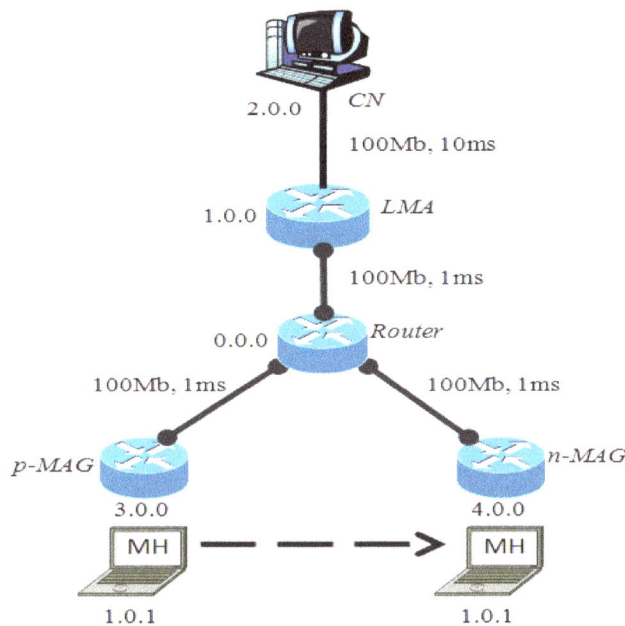

Figure 8. Simulation Scenario Setup of proposed integrated solution of PMIPv6-MIH

In the above simulation scenario, the p-MAG and n-MAG are in separate subnets. The two MAGs have both L_2 and L_3 capabilities that handles handovers. The router is interconnected to the LMA by a series of agents that are organized in a hierarchical tree structure of point-to-point wired links. The router is interconnected to the LMA by a series of agents that are organized in a hierarchical tree structure of point-to-point wired links.

The packet flow of CBR and FTP are simulated and transmitted from the CN to the MH using UDP and TCP. The link delay between the CN and the LMA is set at 10ms while the bandwidth is set at 100Mb. The link delay between the LMA and the respective MAGs is set at 1ms. The CBR and FTP packet size is set at 1000 and 1040 bytes while the interval between successive packets is fixed at 0.001 seconds.

The flow of video traffic is simulated and transmitted from the CN to the MH using myUDP. The video packet size is set at 1028 bytes while the interval between successive packets is also fixed at 0.001 seconds.

5.2. Simulation Results 1

Simulation results for intra-domain traffics are obtained as follows:

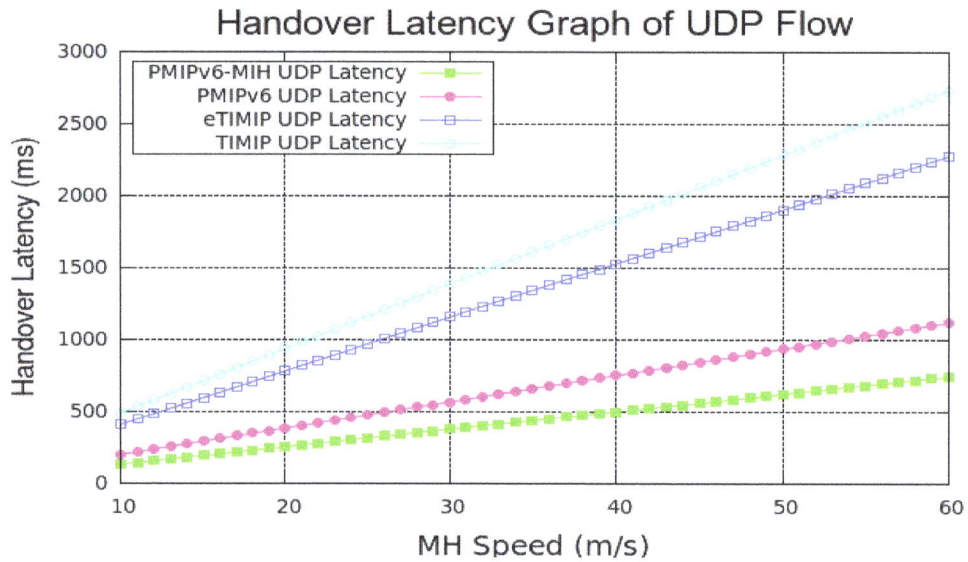

Figure 9. Handover Latency of UDP Flow in micro mobility domain

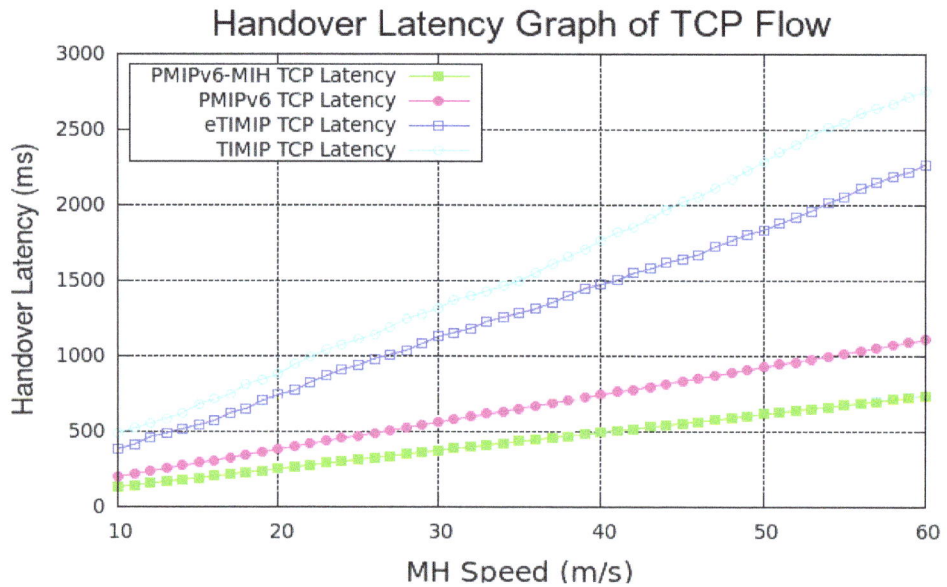

Figure 10. Handover Latency of TCP Flow in micro mobility domain

Packet Delivery Graph of UDP Flow

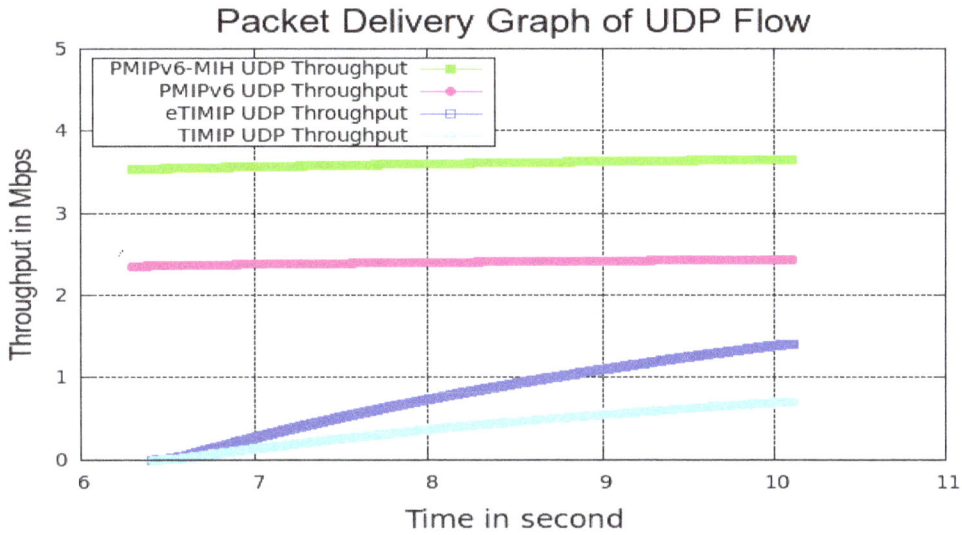

Figure 11. Throughput (Mbps) of UDP Flow in micro mobility domain

Packet Delivery Graph of TCP Flow

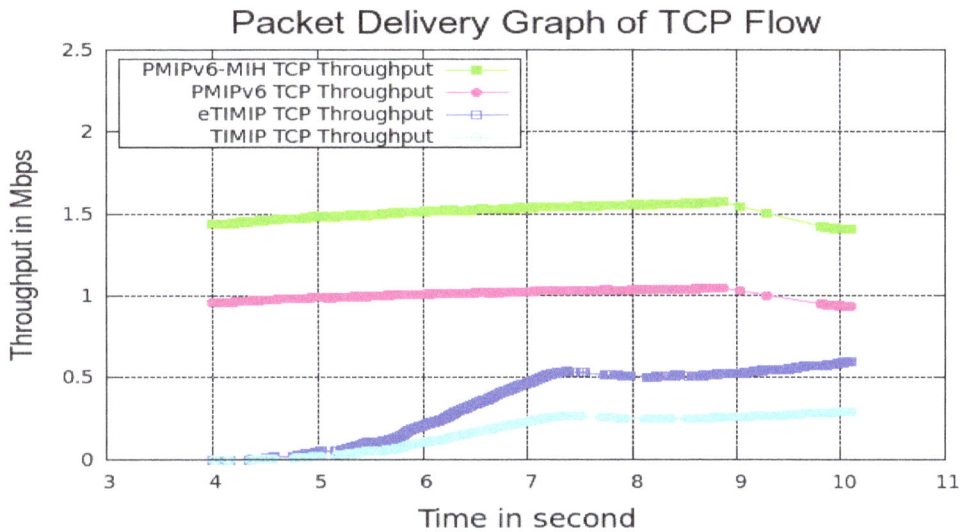

Figure 12. Throughput (Mbps) of TCP Flow in micro mobility domain

As per results shown in figures 9, 10, 11 and 12, it is observed that UDP and TCP performance of eTIMIP and TIMIP increased the handover latency during the MH moves to new network in micro mobility domain. It also noted from the simulation results that performance of throughput also shown degradation. This is due to the fact that, when MH moves away from one network to another in micro mobility domain with high speed mobility, there are lots of operations to perform between the changes of network, such as configuring new CoA, DAD operation, binding registration and MD.

In comparison to PMIPv6, it does not require CoA and DAD as MH is already roaming in the PMIPv6 domain. Once the MH has entered and is roaming inside the PMIPv6 domain, CoA is not relevant since according to the PMIPv6 specification, the MH continues to use the same address configuration. The operation of a DAD is required for a link-local address since address collision is possible between MH, MAG and all MH's attached to the same MAG. The DAD operation may significantly increase handover latency and is a very time consuming procedure. As DAD requires around one second (or even much than one sec.), PMIPv6 introduce a per-MH prefix model in which every MH is assigned a unique HNP. This approach may guarantee address uniqueness. But still PMIPv6 suffers from a lengthy handover latency and packet loss during the handover process when MH speed is high. To overcome these problems, we proposed integrated solution scheme for PMIPv6 that can send the MH-profile to the n-MAG through ND message before handover on L_3 and also reduce the time on L_2 scanning by informing the MH about the channel information of next APs using MIH services.

Based on the proposed solution scheme, the result of handover latency and throughput are better than other mobility protocols. The reason of reduce handover latency and improve throughput in micro mobility domain as below:

❖ The time required to obtain MH profile information can be omitted since n-MAG performs this information retrieval prior to MH's actual attachment.

❖ As the specification of PMIPv6, the time needed to obtain the DAD operation and configure new CoA can also be non-appreciable since n-MAG obtains MH profile and network information through the ND message and performs a pre-DAD procedure like assigning a unique HNP during available resource negotiation with p-MAG and the MH continues to use the same address configuration.

❖ The time required to obtain mobility-related signaling massage exchange during pre-registration may not be considered since this negotiation is established through the ND message before MH attachment. Since the MH is already pre-registered and there is no need to confirm at the n-MAG, therefore the last PBA message send from the LMA may not be considered.

5.3. Simulation Results 2

Simulation results of video transmission over PMIPv6 and PMIPv6-MIH in a micro mobility domain are obtained as follows:

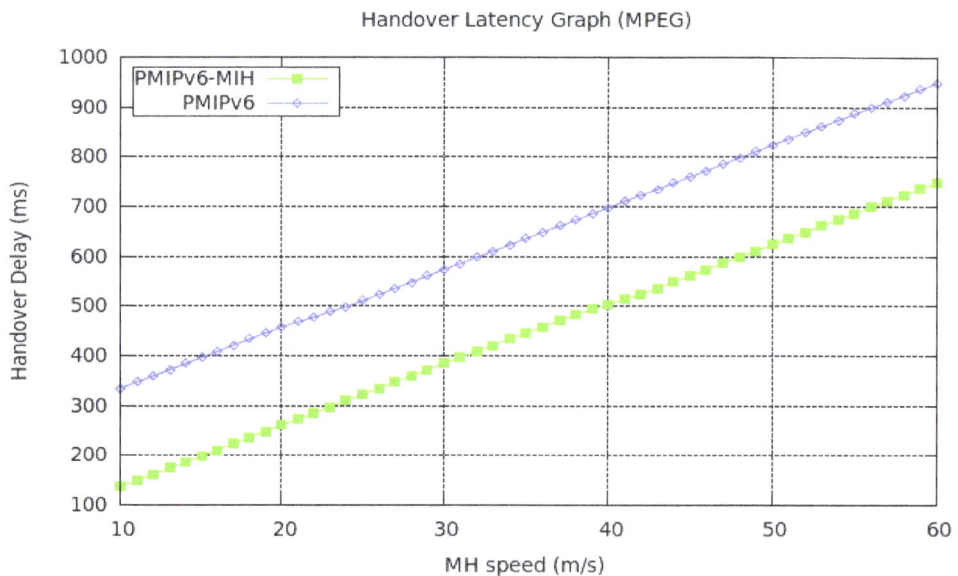

Figure 13. Latency of network-based mobility management protocols

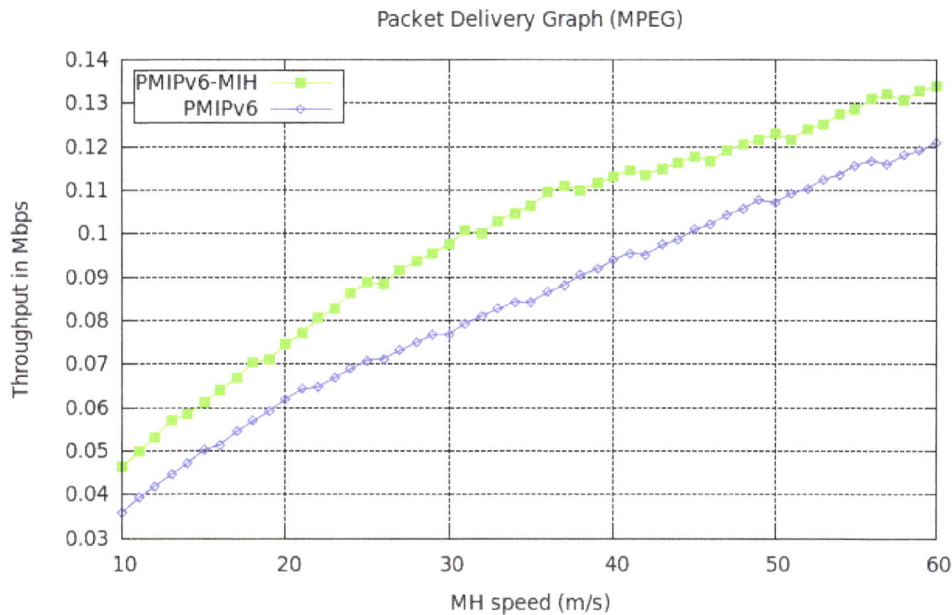

Figure 14. Throughput of network-based mobility management protocols

When the simulation is started, the video start to be transmitted from CN to MH. We can obtain how many frames and packets have been sent or lost after video has been received by MH. The results are as follows:

PMIPv6-MIH
```
Packet sent:p->nA:4651, p->nI:1106, p->nP:1459, p->nB:2085
Packet lost:p->lA:108, p->lI:40, p->lP:25, p->lB:43

Frame sent:f->nA:1331, f->nI:148, f->nP:296, f->nB:886
Frame lost:f->lA:38, f->lI:5, f->lP:8, f->lB:25
```

PMIPv6
```
Packet sent:p->nA:4651, p->nI:1106, p->nP:1459, p->nB:2085
Packet lost:p->lA:268, p->lI:88, p->lP:71, p->lB:109

Frame sent:f->nA:1331, f->nI:148, f->nP:296, f->nB:886
Frame lost:f->lA:91, f->lI:10, f->lP:20, f->lB:61
```

The results of packet and frame are as follows: A denotes number of packets or frames, I, P, B are three different types of frames of MPEG4. The cause of high packet loss, increased in handover latency and decreased throughput of PMIPv6 is due to detection of MHs' detachment and attachment events remain difficult, the need to acquire information from the policy store/AAA server, high number of packets are being transmitted to the p-MAG which cause loss many packets. In order to solve the problems of PMIPv6, the use of our proposed PMIPv6-MIH is expected to reduce lengthy handover latency, decrease the packet loss, increase the throughput and PSNR since the connection to the n-MAG is established before the MH arrived. The time required to obtain MH profile information can be omitted since p-MAG sends all information through the ND message to the n-MAG before MH's actual attachment. Eventually, RS message transmission time may not be appreciable because of the specification of PMIPv6.

The time required to obtain mobility-related signaling message exchange during pre-registration may not be considered since this negotiation is established before MH attachment. Since the MH is already pre-registered and all the necessary information already send to n-MAG through the ND message, therefore no need to confirm at the n-MAG and the last PBA message sent from the LMA may not be considered.

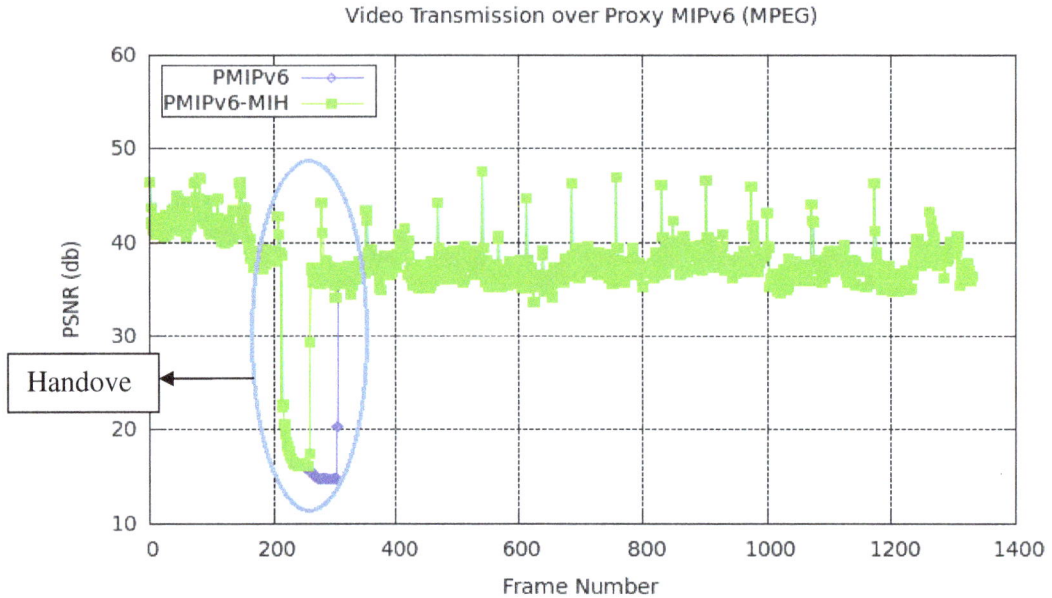

Figure 15. PSNR of network-based mobility management protocols

In figure 15, we can see at the time the PSNR falls when the 250th frame is being transmitted, which is due to the time that handover is taking place during the high speed movement of MH. In addition, the raw video and received video of network-based mobility management protocols can be differentiated with the measurement of computing PSNR [17].

$$PSNR\,(n)_{dB} = 20lg_{10}\,\frac{V_{peak}}{\sqrt{\frac{1}{N_{col}N_{row}}\sum_{i=0}^{N_{col}}\sum_{j=0}^{N_{row}}[Y_S\,(n,i,j)-Y_D\,(n,i,j)]^2}}$$

where $V_{peak} = 2^k-1$, k denotes number of bits per pixels, Y denotes component of source image S and destination image D. PSNR is widely used to compare the video quality of compressed and decompressed images.

The quality of transmitted video images is shown in table below:

Raw Video

Video received over PMIPv6

Video received over PMIPv6-MIH

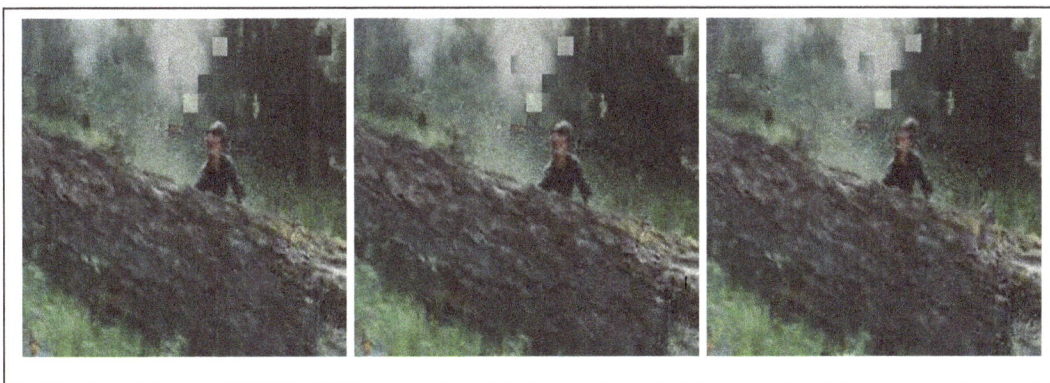

6. CONCLUSION

In this paper, simulations were conducted using ns-2 to evaluate, compare and examine the network-based mobility management protocols and micro mobility protocols with a high speed movement of MH for our proposed integrated solution scheme under intra-domain approaches. As for performance evaluation, we compared performance indicators such as handover latency, packet loss, throughput and video transmission quality. From our analytical analysis and simulation results, we are able to show that our proposed integrated solution of PMIPv6

(PMIPv6-MIH), as a network-based mobility management protocol, performs better than PMIPv6. For our future work, we would like to improve the packet loss, handover latency and also improve the performance of our proposed PMIPv6-MIH for real-time applications in a macro mobility domain with a high speed movement of MH.

REFERENCES

[1]   Johnson, D., Perkins, C., Arkko, J. (2004) "IP Mobility Support in IPv6", RFC 3775, http://www.ietf.org/rfc/rfc3775

[2]   Perkins, Charles.E. (1998) "Mobile Networking Through Mobile IP", In Proceedings of *IEEE Internet Computing*, Vol. 2, Issue 1, pp.58 – 69.

[3]   Yaakob, N. & Anwar, F. (2008) "Seamless Handover Mobility Schemes over High Speed Wireless Environment", In Proceedings of *International Conference on Electrical Engineering and Informatics*, pp. 50-53.

[4]   Yaakob, N., Anwar, F., Suryady, Z. & Abdalla, A.H. (2008) "Investigating Mobile Motion Prediction in Supporting Seamless Handover for High Speed Mobile Node", In Proceedings of *International Conference on Computer and Communication Engineering*, pp. 1260-1263.

[5]   Estrela, P. V., Vazao, T. M. & Nunes, M. S. (2006) "Design and evaluation of eTIMIP – an overlay micro-mobility architecture based on TIMIP", In Proceedings of *International Conference on Wireless and Mobile Communications (ICWMC'06)*, pp. 60-67.

[6]   Estrela, P. V., Vazao, T. M. & Nunes, M. S. (2007) "A Route Optimization scheme for the eTIMIP   micro-mobility protocol" In Proceedings of *IEEE International Symposium on Personal, Indoor and Mobile Radio Communications*, pp. 1-5.

[7]   Kong, K., Lee, W., Han, Y., Shin, .M & You, H. (2008) "Mobility Management for all-IP Mobile Networks: Mobile IPv6 vs. Proxy Mobile IPv6", In Proceedings of *International Conference on Wireless Communications*, pp. 36-45.

[8]   Lee, H., Han, Y. & Min, S. (2010) "Network Mobility Support Scheme on PMIPv6 Networks", *International Journal of Computer Networks & Communications (IJCNC)*, Vol.2, No.5

[9]   NS-2 home page, http://www.isi.edu/nsnam/ns

[10]   Peak Signal-to-Noise Ratio (PSNR), http://en.wikipedia.org/wiki/Peak_signal-to-noise_ratio

[11]   Kwon, T.T., Gerla, M. & Das, S. (2002) "Mobility management for VoIP Service: Mobile IP vs. SIP", In Proceedings of *IEEE Wireless Communications*, Vol. 9, Issue 5, pp. 66 - 75.

[12]   Jung, J., Kim, Y. & Kahng, H. (2004) "SCTP Mobility Highly Coupled with Mobile IP", In Proceedings of *Telecommunications and Networking -- 11th International Conference on Telecommunications*, Issue 3124, pp. 671-677.

[13]   Columbia IP micro-mobility suite (CIMS), http://tagus.inesc-id.pt/~pestrela/ns2/mobility.html

[14]   Abdalla Hashim, A.H., Ridzuan, F. & Rusli, N. (2005) "Evaluation of Handover Latency in Intra-Domain Mobility", *Journal of World Academy of Science, Engineering and Technology*

[15]   Media Independent Handover (MIH), http://en.wikipedia.org/wiki/Media-independent_handover

[16]   Taniuchi, K., Ohba, Y., Fajardo, V. (2009) "IEEE 802.21: Media Independent Handover: Features, Applicability, and Realization", In Proceedings of *IEEE In Communications Magazine*, Vol. 47, Issue: 1, pp. 112–120.

[17]   Ke, C. An example of multimedia transmission over a wireless network. http://hpds.ee.ncku.edu.tw/~smallko/ns2/Evalvid_in_NS2.htm

7

Achieving Transmission Fairness in Distributed Medium Access Wireless Mesh Networks: Design Challenges, Guidelines and Future Directions

Salitha Priyanka Undugodage and Nurul I Sarkar

School of Computing and Mathematical Sciences
Auckland University of Technology, Auckland, New Zealand

Nurul.sarkar@aut.ac.nz

ABSTRACT

Wireless mesh networking gained an international interest over the years as a result to high recognition in the wireless industry as a cost effective, scalable, wider coverage and capacity capable wireless technology. The contention based distributed medium access in wireless networks has advanced not only in supporting the quality of multimedia but also achieving high throughput and to minimize packet delay overheads in legacy systems. Unfortunately, the impact of such enhancement has not been fully justified with mesh network environments yet. The medium access frames are required to be contended over multi-hops to overcome the challenges of improving overall system performance through concurrent transmissions. The goal of this paper is to discuss the issues and challenges of transmission fairness and the effect of concurrent transmission on system performance. To mitigate transmission fairness issues, we review existing open literature on mesh networking and provide guidelines for better system design and deployment. Finally, we conclude the paper with future research directions. This study may help network designer and planner to overcome the remaining challenging issues in the design and deployment of WMNs worldwide.

Keywords
Wireless Mesh Networks (WMN), Frame Aggregation, Block Acknowledgement (BA), Reverse Direction Grant (RDG), Carrier Sensed Threshold (CST)

1. INTRODUCTION

The Wireless Mesh Network (WMN) is a set of wireless nodes where each node can communicate directly with one or more peer nodes. WMN has been standardized by IEEE 802.11 Task Group "s" to develop a set of standards for WMNs under the IEEE 802.11s. Further the IETF (Internet Engineering Task Force) had also setup wireless mesh networking called Mobile Ad-Hoc Network (MANET) with a separate set of standards. Both MANETs and WMN nodes exploit the redundancy of connected nodes and have the ability of self-organize, self-discover, self-heal, and self-configure. However, in real-world applications, MANETs are implemented with mobile and more power constrained nodes, and the infrastructure is less self-organized. In contrast, WMNs are typically a collection of more organized stationary nodes and may use multiple radios for the purpose of wireless mesh backhauling for Wireless Local Area Network (WLAN) with one radio and the other radio for Access Point (AP) functionality [1]. Although WMNs could extend the wireless coverage as a cost-effective backhaul solutions it has many challenges, especially when increasing the per user data rate of multiple concurrent sessions between multi-hop mesh nodes in serving as backhaul WLAN technologies. These challenges are as a result of 802.11's shared medium access constrains in achieving transmission fairness, especially in multi-hop networks.

In this paper we address some of the key issues of such constrains and provide guidelines for network researchers and designer for efficient system design and deployment of such system. The remainder of this paper is organized as follows. In Section 2 we highlight 802.11 Physical layer (PHY) and Medium Access Control (MAC) layer standards in WMNs focusing on distributed medium access protocols. The issues and challenges in designing WMNs are also discussed. Section 3 presents WMN architecture highlighting the transmission fairness issues in a multi-hop contention based shared medium access. In Section 4, we discuss MAC enhancements for multi-hop WMNs medium access efficiency. Section 5 discusses transmission fairness focusing on optimum concurrent transmission in a mesh network. An amendment to the shared MAC with a reverse direction MAC frame pull mechanism to optimize concurrent transmission is also discussed. In Section 6, we present guidelines for WMN design and deployment and future research directions. Finally, a brief discussion in Section 7 concludes the paper.

2. WMN DESIGN CHALLENGES IN DISTRIBUTED MEDIUM ACCESS

One of the primary objectives of 802.11s WMN standardization was to define the 802.11 PHY and MAC layers to create a Wireless Distribution System (DS) which is capable of automating topology learning and wireless path configuration for self-learning, self-forming and self-healing wireless paths. The standard defines dynamic and radio-aware path selection mechanism to delivery of data on both single-hop and multi-hop networks. Any wireless node complying with these functionalities are said to be wireless mesh capable nodes which forms a WMN or a mesh cloud. One of the key issues in WMN standardization is the adaptation of legacy distributed medium access schemes to share the medium which has inherent unfairness in achieving concurrent transmissions between mesh nodes in a multi-hop mesh network. However, it is important that WMN standards should address these challenging issues without compromising the compatibilities of WMNs to continue to evolve as a cost-effective backhauling technology for WLANs [2] [3] [4].

2.1 Mesh Network PHY and MAC layer Standards

The IEEE 802.11 PHY and MAC Layer standards were first introduced in 1997. Since then multiple standards had evolved under different IEEE Task Groups as "a" (TGa) and "b" (TGb) in 1999, "g" (TGg) in 2003 and "n" (TGn) 2007. These WLAN standards had evolved with the number of enhancements into the PHY and MAC layers mainly to improve raw data speed and propagation range while maintaining backward compatibility with the previous standards. Consequently, the 802.11g APs are backward compatible in connecting 802.11b Stations (STAs). Similarly IEEE 802.11n APs are backward compatible in communicating to 802.11a/b/g STAs. A WLAN operating in multi-mode supporting more than one mode is said to be in "mixed mode" whereas a WLAN is said to be operating in "Green field" if all STAs only support native highest performing mode. The most capabilities of Green field operation are compromised when operating in mixed mode. In theory 802.11s could operate on any 802.11 PHY layer standard supporting either mixed mode or green field but it is sensible for all mesh nodes to be deployed in a same mode (e.g. green field network) for greater performance [5].

Figure 1 shows 802.11 MAC protocol structure. The MAC layer defines the data link between two mesh nodes and exchanges MAC Service Data Units (MSDUs) packed into MAC Protocol Data Units (MPDU) and carried over the PHY Protocol Data Unit (PPDU) as per the original 802.11 MAC standards. The main concern observed in a wireless mesh is that the standard requires every successfully non multicast and broadcast frames received at each mesh node to be acknowledged causing considerable packet delays in multi-hop communications (a frame needs to cross multiple hops in reaching the destination) [6] [7].

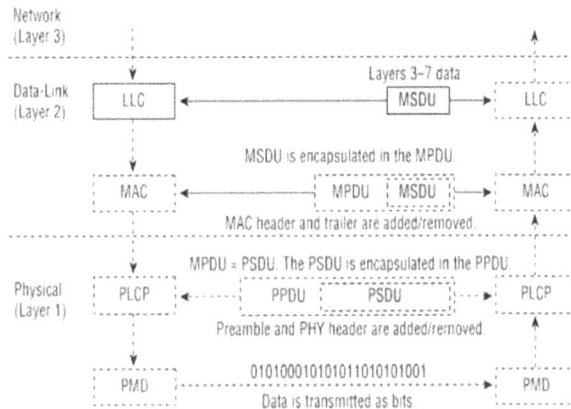

Figure1: Media access control (MAC) protocol structure.

2.2 WMN MAC Layer Protocol Design Challenges

IEEE 802.11 medium access protocol is based on Carrier Sense Multiple Access with Collision Avoidance (CSMA/CA) to avoid frame collisions in a shared wireless channel. This medium access layer is similar to IEEE 802.3 wire-line medium access which is based on Carrier Sense Multiple Access with Collision Detection (CSMA/CD) where medium access coordination mechanism is to detect rather than avoid collisions only. CSMA/CD is not suitable for wireless network because network interface cards cannot transmit and listen on the same wireless channel simultaneously. The Receiver (Rx) must receive the incoming frame fully before the wireless interface could switch from "receive" mode to "transmit" mode to transmit a frames which makes CSMA/CA ideal in wireless contention based shared access. The idea of carrier sense is listen before transmit in assessing nearby node engage in transmission. The Transmitter (Tx) will refrain from transmission if received energy level at a Tx at any time slots duration higher than a fixed Carrier Sensed Threshold (CST) to avoid collision [8] [9]. Although this is acceptable in an AP centric WLAN implementation, it is a major concern in WMNs where exposed mesh nodes suspend any concurrent transmission to avoid collisions. To overcome this problem it is possible to separate mesh nodes so that they do not necessarily exposed to each other. However this may lead to increased collisions at the Rx end if a Transmit mesh node estimates a lower energy level from another hidden mesh node that could transmit at the same time slot. This is known as Hidden Station Problem which is a well-known issue in CSMA/CA medium access. The IEEE has standardized the 802.11 Distributed Coordination Function (DCF) for contention based medium access minimizing the hidden station problem by either 2-Way handshaking where each

MAC frame is acknowledged by an acknowledge frame (ACK) or 4-Way handshaking with an additional contention free medium protection called RTS/CTS handshaking or both by providing any hidden nodes access the shared channel. In fact, as shown in Figure 2 RTS/CTS is a virtual carrier sense mechanism to block any attempt to transmit by any exposed nodes for a specified duration called Network Allocation Vector (NAV) exclusively allocating the channel to the nodes that raise the RTS and CTS as shown in Figure 2. However, a complete elimination of hidden mesh nodes as well as exposing all mesh nodes in a WMN refraining concurrent transmission between mesh nodes could be challenging, as it leads to unfairness in sharing channel for multi-hop transmissions [10]. The multi-hop network throughput scenario/analysis is discussed next.

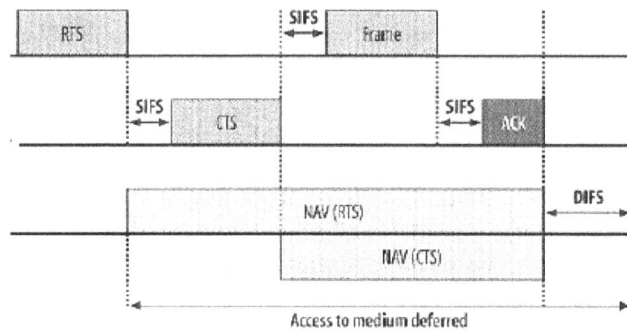

Figure 2: RTS/CTS and NAV timing diagram.

2.3 Multi-Hop Network Throughput Analysis

To eliminate the hidden station problem and to avoid collisions, mesh nodes must maintain received power levels within carrier sensed threshold (CST). However this could prevent the nodes that are exposed utilizing the medium for concurrent transmission due to the DCF contention access scheme which is called Exposed Station Problem which is a major barrier to exploit concurrent transmissions in multi-hop mesh networks. Inability to perform concurrent transmission between mesh nodes in a multi-hop network would increase the MAC frame transmission waiting times with the increase of mesh node density resulting in rapid throughput degradation. In other words, the balance between hidden and exposed nodes is crucial in optimizing the concurrent communications in a WMN [11] [12] [13].

When all mesh nodes are in the same collision domain and if N numbers of nodes are exposed to each other, the probability of successful frame transmission would be 1/N. Assuming N number of hops or N +1 number of nodes in an exposed collision domain and no packets are losses at relay nodes between source to destination as well as negligible propagation time between nodes then the single-hop normalized end to end throughput would be:

$$\frac{1}{N + (N-1)\frac{T_i}{T_p}}$$

Where Tp = Time to Transmit / Receive payload at a node

   Ti = Intermediate relay node transmission latency

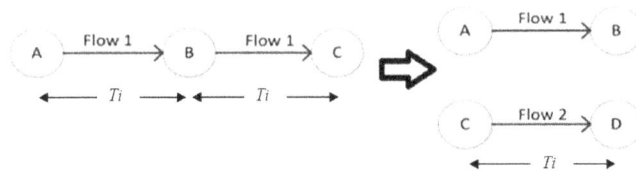

Figure 3: Multi hop concurrent transmission delay

The above expression shows that the packet delay overheads could be minimized to improve the raw data transmission efficiency by minimizing Ti which will in return increase the throughput between mesh nodes. Applying this model to the concurrent transmission scenario as shown in Figure 3 for a contention based shared medium access scheme where Tp is unbounded, indicating that constrains in concurrent transmission between multi-hops could limit the per user throughput when multiple user consume network bandwidth [14] [15] [16] [17].

2.4 IEEE 802.11 DCF Access Mechanism

In 2-way handshaking (Figure 4) when the medium is idle and the nodes contending for the medium will access the channel immediately after the period of Short Inter-Frame Space (SIFS). If expecting any acknowledgement frames for prior transmissions and wait further duration up to DCF Inter-Frame Space (DIFS).

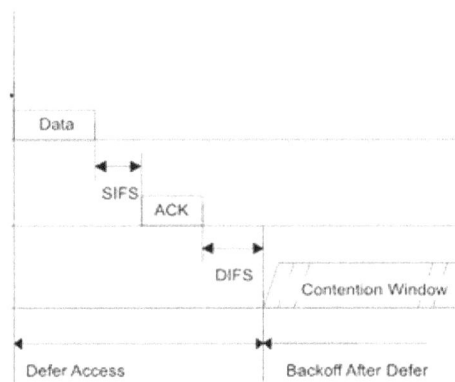

Figure 4: Inter-frame spacing and back-off.

If the medium is not idle, nodes will continue to wait a random back-off period set up in the Back-off Counter (BC). The node transmits when BC expires to minimize any possible collisions in transmissions. BC is a uniformly distributed random number between 0 and a Contention Window (CW) defined. The CW size is initially assigned CW_{min}, and increases by doubling CW with an upper bound of CW_{max} when collision is experienced but every successful transmission will reset CW back to CW_{min}. CW size is measured in terms of slot time which is defined for different 802.11 PHY standards [12] [2]. Figure 4 illustrates the basic concept of inter-frame spacing and back-off mechanisms.

These inter-frame spacing and random back-off introduce delay overheads where medium left unutilized before each transmission but built into the DCF scheme to minimize possible collisions [9]. Figure 5 illustrates the overheads associated with 2-way handshaking. This delay overhead is even worst with 4-Way handshaking where additional RTS/CTS NAV delays further contribute to longer underutilised medium assuring the medium protection for transmission without collision at a cost of overall raw data rate degradation. In fact 4-Way handshaking or RTS/CTS is usually recommended to be used only when long frames are to be transmitted where a retransmission degrades system performance in case of a frame losses due to collisions [12].

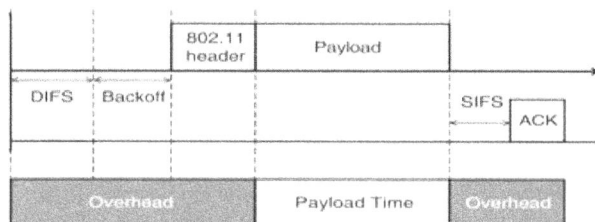

Figure 5: Overheads in 2-way handshaking.

3. WIRELESS MESH NETWORK ELEMENTS

Figure 6 shows the architecture of a typical wireless mesh network. The 802.11 standard defines the WLAN Basic Service Set (BSS) where a set of WLAN STAs that are associated to an AP or each other in an ad hoc manner. Similarly in a WMN, the mesh nodes called as Mesh Point (MP) are associated to each other based on the 802.11s standardized Mesh Basic Service Set (MBSS). In other words the MBSS is a set of MPs that are associated to each other forming a transparent single broadcast domain mesh cloud. However, unlike WLAN BSS STAs the MPs in a MBSS has the relaying capability and MPs could exchange MAC frames over multiple wireless hops by maintaining established mesh links with peering MPs in its neighborhood. The MBSS mesh topology formed by MPs searches for potential MPs present in the neighbourhood by either active scanning or passive listening over air waves and exchanging the Mesh Profile which consisting of a Mesh ID, Path selection protocol identifier, and Link metric identifier. In fact, the Mesh Profile that matches each other got associated them-self forming partial or a full mesh topology. Once associated the MPs establish mesh links and continue to exchange beacons frames for topology maintained and concatenated set of mesh links established via reachable MPs maintained mesh paths in a mesh topology [7] [6].

3.1 WMN Mesh Functionality and Routing

MPs in a WMN could have one or multiple optional functions other than the mandatory mesh function, such as the AP function which allows an MP to function as an AP to connect 802.11 WLAN STAs and such a mesh node is called a Mesh Access Point (MAP). A MP that could translate 802.11s MAC frames to 802.11 WLAN MAC frames is called a Mesh Gateway (MG). Having gateway functionality and an MG may have external gateway functionality as well to connect an MP to an external 802.3 LAN or wired backhaul such an MP is called a Mesh Portal Point (MPP) as illustrated in Figure 6.

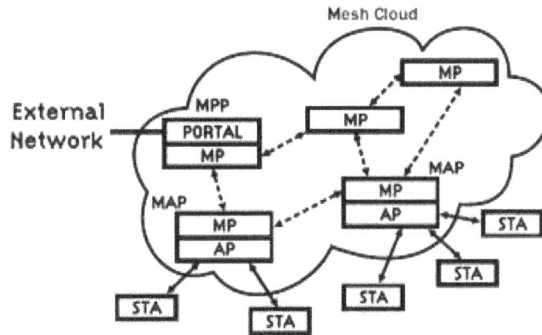

Figure 6: Wireless mesh network topology.

The MPs learn the mesh topology through routing protocols and an interesting feature in 802.11s WMN standard is that the definition of its own routing protocols for frame forwarding and path selection in the MAC layer itself without depending on network layer or usual TCP/IP routing Protocols.

The IETF routing and forwarding standard for MANET called Hybrid Wireless Mesh Protocol (HWMP) which provides both on demand routing with Ad-hoc On-demand Distance Vector (AODV) and proactive tree-based routing with Optimized Link State Routing (OLSR) is used in 802.11s WMNs as well. Although WMN framework allows multiple routing protocols to be implemented in a MANET and only one of them could be active in a Mesh cloud [8] [4].

3.2 Spatial Bias Multi-Hops WMN

In a multi-hop mesh topology, a user performance depends on the number of hops the frame had to travel in reaching the destination. Higher the number of hop counts, the lower the overall throughput achieved due to contention overheads at each hope resulting unfairness in spatial resources use for an MP which has higher number of hops to the destination. The scenario called spatial bias where more the mesh hops in a mesh path, higher the frames affected which is scalability concern in designing WMN. Research has shown that the bandwidth starvation due to spatial bias in multi-hop could be optimized by dynamically adjusting the packet size and the minimum contention period based on congestion experienced due to spatial bias [6].

4. MAC ENHANCEMENT SCHEMES

Figure 7 shows the access control mechanism based on arbitration inter-frame space (AIFS) in EDCA. The 802.11standards had not considered priority base Quality of Service (QoS) and capable of serving only best effort delivery data over WLANs. But with the increasing demand for carrying multimedia traffic over WLANs, the IEEE to form separate Task Group "e" (TGe) introducing standards for Wireless Multimedia (WMM) under the 802.11e standard which extend DCF with QoS capabilities. In the 802.11e standard, different traffic types are classified based on 8 different priority values mapped onto the 4 FIFO queues, called Access Classes (ACs) where each AC behaves like a virtual node. This WMM medium control coordination scheme is called Enhanced Distributed Channel Access (EDCA) and the contention time DIFS is defined for each corresponding traffic AC as the AIFS as illustrated in Figure 7 [18] [13].

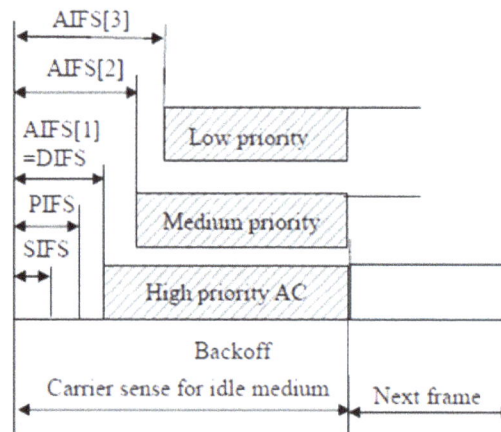

Figure 7: Access control based AIFS in EDCA.

Higher priority traffic category will have a shorter AIFS than a lower priority traffic category which means lower priority traffic must wait longer time than high priority traffic before accessing the medium. Although the probabilistic priority mechanism for allocating bandwidth based on traffic categories has no guarantees of delivery between MPs in a mesh cloud, the EDCA is the mandatory medium access scheme in 802.11s WMNs. As a result 802.11s standard specifies another medium access scheme called Mesh Deterministic Access (MDA) as an option in the Mesh Coordination Function (MCF). MDA is also a distributed and reservation based deterministic medium access scheme and capable of providing prioritized QoS with delivery guarantees. The advantage of MDA compared to EDCA is that the mesh nodes could negotiate a periodic transmission opportunity for collision free transmissions. However, MDA capable mesh nodes need to be synchronized each other and therefore it becomes more complex due to the ad hoc nature of the mesh topology. Further in 802.11s based WMNs, synchronization is optional due to its distributed nature and not all mesh nodes are required to participate in the MDA scheme which could impact the presence of contention from non-MDA mesh nodes in the neighbourhood [19] [17] [20]. The various MAC enhancement mechanisms for WMNs are discussed next.

4.1 Transmission Opportunity and Frame Aggregation

IEEE 802.11e WMM standard also adds additional MAC enhancements such as aggregating frames to be transmitted during the opportunity gained by contention scheme which is named Transmission Opportunity (TXOP) and Block Acknowledgement (BA). It enables the receiver to acknowledge the successful reception of multiple frames using a single BA frame. TXOP is a bounded time interval defined by a maximum duration in which a series of frames are transmitted. TXOP Limit, which depends on the AC, is the maximum time a node could hold a channel after a successful contention. Frame aggregation concept allows Aggregated MAC Service Data Unit (A-MSDU) to be sent to the same receiver concatenated into a single MPDU and transmitted either when transmit queue reaches the maximal A-MSDU threshold or any frame timeout condition. The BA contains a bitmap to selectively ACK individual frames in an aggregated frame burst allowing a block of frames separated by an inter frame spacing of SIFS with same AC to be transmitted without waiting for acknowledgment. A-MSDU transmitted is followed by a BA Request (BAR) frame to enquire which frames have been received successfully which is answered with a BA frame for every successful frame delivery [5] [21] [22] [23].

4.2 IEEE 802.11n MAC Enhancements

IEEE 802.11n WLAN standard added further enhancement for the frame aggregation with another level of aggregation called Aggregated MAC Protocol Data Unit (A-MPDU), aggregating MPDU sub frames to a single PHY frame. Unlike A-MSDU there is no waiting time for an A-MPDU and the number of MPDUs aggregated depends on the number of frames in the transmit queue at the time of gaining the TXOP. MSDUs within an A-MSDU are addressed to the same receiver whereas MPDUs within an A-MPDU need not be to the same receiver. A blend of both A-MSDU and A-MPDU over two stages will maximize throughput efficiency. Further BAR is made optional and Rx could respond with BA after each aggregated frame without waiting for a BAR which removes the BAR overhead and eliminate the possibility of retransmission. This could be any failure to receive BAR and allows multiple aggregated frames to be acknowledged by a single BA. BA could be either expected immediately as a response to the BAR or could be a delayed BA [24] [25] [26].

Another medium access enhancement introduced in 802.11n is the reduced and zero inter-frame spacing (RIFS and ZIFS) to minimize the overhead between frames. Inter frame spacing is required within TXOP between frames and between the last frame and BAR. This is reduced from SIFS to RIFS where (RIFS << SIFS) between multiple aggregated frames or completely removed which means RIFS = ZIFS eliminating the overhead due to inter-frame spacing resulting more bit transmission using TXOP [5] [27] [28].

4.3 Reverse Direction (RD) Flow

IEEE 802.11n standard has also been enhanced frame aggregation called Reverse Direction (RD) flow, which improves the TXOP effectiveness by allowing frame transfer from responder to the originator during originator's TXOP. RD flow initiates with RTS/CTS handshake and the peers make a request inside the RTS/CTS NAV duration. RD flow requires the TXOP originator to grant permission to the responder to send data frames aggregated in the reverse direction while being responsible for channel ownership. Gain in throughput performance would be achieved in RD flow by granting responder node to transfer frames without contention related overheads [29] [30].

4.4 Green Field High Throughput Mode

IEEE 802.11n standard is backward compatible with previous generations 802.11a/b/g and operates in three modes, namely Legacy Mode, Mixed Mode and Green Field Mode. In Legacy mode, frames are transmitted in the legacy 802.11a/g MAC format frames with no 802.11n MAC features. In the Mixed Mode, 802.11a/g frames are transmitted with a preamble compatible with the legacy 802.11a/g such that it can be decoded by legacy 802.11a/g devices while transmitting 802.11n frames with an initial training sequence format which occupies less air time to reduce per-transmission overheads. Therefore, medium protection RTS/CTS handshaking is required to permit communication with legacy stations to ensure legacy devices sense the channel busy state. Thus, 802.11n devices have to pay significant throughput penalties when legacy devices are served in mixed mode. The Green Field mode is exclusively for 802.11n devices only with high throughput (HT) format preamble is used in MAC frames for HT transmission. If no legacy devices served the 802.11n WLANs operate in maximum HT performance then it is said to be a green field network [28] [31] [32] [33].

5. TRANSMISSION FAIRNESS STRATEGIES

The DCF and the enhanced EDCA contention based distributed access schemes have been successful in all 802.11 standards irrespective of other contention free schemes used in the standards to avoid contention in 802.11n [34].

In contention based distributed medium access schemes for multi-hop networks, nodes are required to accommodate multiple concurrent transmissions. It is important to distribute/separate the nodes to multiple collision domains so that the nodes can be exposed in that collision domain. To carry frames across the collision domains, an overlap between collision domains is required where a node within overlapping area would be the transit node for interconnecting two or more collision domains [35] [36] [37].

Table 1 compares the four main distributed medium access mechanisms for WMNs. The comparison is based on various factors, including IEEE standards, MAC type, operating mode, frame aggregation, support for block Ack (BA), reverse direction (RD), and NAV, and priority.

Table 1: Comparison of distributed medium access schemes

Scheme	DCF	EDCA	MDA	HT EDCA
IEEE Standard	802.11a/b/g	802.11e	802.11s	802.11n
MAC type	Contention	Contention	No contention	Contention
Mixed mode supported	Yes	Yes	Yes	Yes
Frame aggregation	No	Yes	Yes	Yes
Support BA	No	Yes	Yes	Yes
Support for Reverse Direction flow	No	No	No	Yes
RIFS/ZIFS	No	No	No	Yes
Support NAV Protection	Yes	Only Long frames	Yes	Mixed mode
Synchronization	Optional	Optional	Mandatory	Optional
AC Priority QoS	No	Yes	Yes	Yes
Exposed STA avoid scheme	No	No	Yes	No
Concurrent Tx scheme for MPs	No	No	No	No

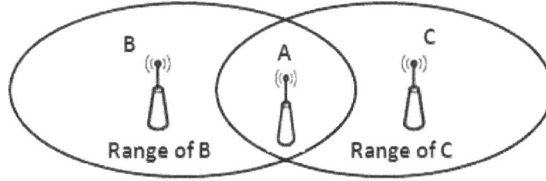

Figure 8: Transit mesh point exposed areas.

A transit node will be exposed to multiple collision domains to contend in both collision domains to have longer span of time for a transmission as shown in Figure 8. This scenario is called "Neighborhood capture problem" where the transit node will hardly find the free medium to access both collision domains [38] [39] [40] [41].

5.1 Channel Estimated Power Management

In contention based medium access, the frames are lost as a result of collisions or transmission errors. The transmission errors occurred due to poor channel conditions. The dynamic link adaptation using modulation and coding scheme (MCS) and forward error correction (FEC) compromises the raw data rate by mitigating transmission errors. However, frame losses are normally occurred due to collision at the receiving end when a hidden node attempts to transmit data as illustrated in Figure 9. In a WMN setting, the Tx power and CST levels at each node is a decisive factor in fixing the propagation range of exposed MPs. To select the optimum CST level one could avoid hidden MPs as well as limit the exposed MPs [42] [43] [44].

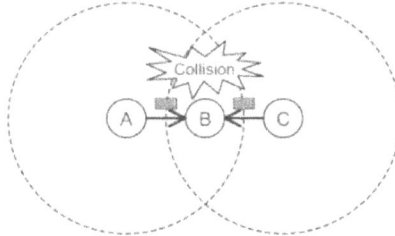

Figure 9: Transit node receiver end collision.

To optimize concurrent transmission between MPs, a strict power management and channel condition estimation at each MP in WMNs is required to ensure MPs are exposed to well manage collision domains.

To analyse MP power management, let us look at the well-known formula for Free Space Power Loss P_L is given by

$$P_L(dB) = 20Log_{10}(d) + 20Log_{10}(f) + 32.45$$

$$P_L(dB) = P_{Out}(dB) + RSSI(dBm)$$

Where d is the distance (in km) between the nodes, f is the signal frequency in MHz and P_{Out} is the MP Transmit power and the RSSI is the receiving MP Received Signal Strength Indicator in dBm.

Even at low RSSI levels due to poor channel conditions or high P_L the MPs could establish a transmission by adopting low bit rate MSC schemes. However, to avoid two MPs exposed to each other it is required to lower the transmit power P_L to a level such that RSSI level at the receiver is insufficient to establish any transmission even with lowest possible MSC scheme [45] [38].

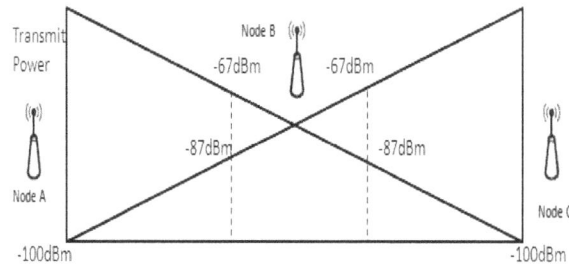

Figure 10: Transmit power management.

Figure 10 demonstrates the transmit power management strategy in WMNs. In 802.11, every frame transmitted is expected to be acknowledged when delivered. If the frame is not acknowledged within a predefine timeout duration, the frame is considered to be lost. The reason for acknowledgement failure which could be either a transmit frame lost due to poor channel conditions or a frame collision that is not known by the sender. However with the introduction of 802.11e MAC enhancements, the BA could be used by the sender to assess the exact reason for an unsuccessful frames delivery. If return indicate many error frames transmitted in the Bitmap, the channel suffers from poor channel conditions. Further if BA is never returned during the BA timeout period then frames could have been collided. This clarity on transmit frames is useful in a WMN to adjust the transmit power levels and CST of an MP to avoid collisions as a result of an optimum collision domain separation [30] [46] [11] [33].

5.2 Design of RD Pull Collision Avoidance Scheme

WMN in Greenfield mode will ensure all MPs support 802.11e MAC layer enhancements as well as 802.11n HT features consistently. In a Greenfield mode, a WMN can be implemented in multiple collision domains where each collision domain is interconnected to the neighboring collision domains through one or more transit MPs. This strategy would allow concurrent transmissions without interfering nodes using contention based medium access mechanisms in neighboring collision domains. However, this may lead to transit MP starvation due to neighborhood capture problems. We suggest that the reverse direction (RD) pull mechanism can be used to avoid collisions due to neighborhood capture problems (discuss below). This strategy can be used without modifying the EDCA contention based medium access scheme in WMNs, especially when modeling using a credible simulation package, such as OPNET or ns-2 for performance evaluation [47] [48]. The RD algorithm is described below.

- Neighborhood capture most transit MPs in receiving state and request for RD flow from the sender during sender's TXOP using either CTS frame or BA frame.
- The reverse direction grant (RDG) request could be initiated by the neighborhood captured MPs when its transmit buffer reaches threshold limit. The system may experience collisions when contending for the shared channel or any higher layer strict delivery conditions.

- Receiver's RDG request is responded by the TXOP's own MP with a RDG allocating any excess TXOP to pull the MAC frames in RD. If the excess TXOP is insufficient to pull all MAC frames waiting to be transmitted, another RDG request can be processed before the end of TXOP.
- If an MP had received a RDG request it will contend to the shared channel to respond with a RDG in order to pull frame in RD.

Although the RD pull mechanism may halt concurrent transmissions, addressing the neighbourhood captured problems will in fact optimize the concurrent transmission in WMNs [14] [47]. This is an important strategy in achieving high throughput in WMNs. The network design guidelines to improve system performance are discussed next.

6. DESIGN GUIDELINES AND DISCUSSION

The three design guidelines for optimizing multi-hop WMNs using concurrent transmissions are discussed below.

(1) **Network design using Green field mode**: It is important to operate all mesh network MPs in green field mode only. Use all 802.11e wireless multimedia (WMM) and 802.11n HT features to maximize system performance as well as capitalise on novel features such a reverse direction (RD) flow.

(2) **Network design by splitting:** Split the WMN into multiple collision domains in such a way that MPs in each collision domain can transmit frames within the domain independently to exploit concurrent transmissions. This could be done by varying the transmitter power and CST so that MPs in different collision domains do not associate each other even at the lowest possible modulation and coding scheme (MCS). This strategy of WMN design will optimize network throughput performance.

(3) **Network design by exposed collision domain:** Ensure that at least two MPs can be exposed to any two collision domains to transit traffic between the two collision domains. This strategy will avoid a single point of failure of a single MP. However, the impact on such a transit MP due to neighbourhood capture problem for being exposed to more than one collision domains needs to be addressed. An effective solution would be to use RD pull mechanism that could pull traffic from transit MPs by granting the RD flow to achieve transmission fairness and optimising concurrent transmissions.

There are many challenging issues in the design, implementation, and deployment of WMNs. Some of the issues are discussed in Section 2 of this paper. Now the question may arise about the better ways of designing WMNs for optimum performance. However, high performance WMN can be (re)design if some obvious guidelines were adopted. In this section we formulated three guidelines for network designers and researchers for optimizing multi-hop WMNs. First, network should be designed using Green field mode only. This means that WMN design using high performance 802.11n devices only. Second, network should be designed by splitting a WMN into multiple collision domains to exploit concurrent transmissions. Third, network should be designed by exposed collision domains. This means that at least two MPs can be exposed to any two collision domains for better system performance.

7. CONCLUSION AND FUTURE WORK

The existing distributed medium access schemes and MAC-layer enhancements for improving concurrent transmission among mesh points (MPs) in WMNs are discussed. The evolution of the 802.11 standards PHY and MAC layers with the assumption that contention based distributed medium access protocols will continue to be the most accepted mechanisms in sharing the medium not only in WLANs but also in multi-hop WMNs. We reviewed existing mechanisms for improving the performance of a typical WMN by concurrent transmission among MPs. These mechanisms can be added to the 802.11 standards especially transmit opportunity (TXOP), frame aggregation, block acknowledgement, reduce inter frame spacing and reverse direction flow. Based on the findings from open literature we identify a high throughput green field WMN separated into multiple collision domains would be the best design strategy to optimize concurrent transmissions in WMNs. We also suggest that reverse direction (RD) pull mechanism can be used to avoid neighbourhood capture issues at a MP specially when handling transit traffic between collision domains. Development of an extensive simulation model of a large WMN with multiple transit mesh nodes handling contention and concurrent transmission is suggested as future work.

Acknowledgment

This research was supported in part by the Faculty of Design & Creative Technologies under Summer Studentship grant 2012-2013, Auckland University of Technology.

REFERENCES

[1]   G. R. Hiertz, Z. Yunpeng, S. Max, T. Junge, E. Weiss, and B. Wolz, "IEEE 802.11s: WLAN mesh standardization and high performance extensions," *IEEE Network*, vol. 22, no. 3, pp. 12-19, 2008.

[2]   G. Hiertz, D. Denteneer, L. Stibor, Y. Zang, X. P. Costa, and B. Walke, "The IEEE 802.11 Universe," *IEEE Communications Magazine*, vol. 48, no. 1, pp. 62-70, 2010.

[3]   S. G. F. Kaabi, F. Filali, "Channel Allocation and Routing in Wireless Mesh Networks: A survey and qualitative comparison between schemes," *International Journal of Wireless and Mobile Nteworks (IJWMN)*, vol. 2, no. 1, pp., 2010.

[4]   G. R. Hiertz, D. Denteneer, S. Max, R. Taori, J. Cardona, L. Berlemann, and B. Walke, "IEEE 802.11s: The WLAN Mesh Standard," *IEEE Wireless Communications*, vol. 17, no. 1, pp. 104-111, 2010.

[5]   M. A. García, M. Angeles Santos, and J. Villalón, "IEEE 802.11n MAC Mechanisms for High Throughput: a Performance Evaluation," presented at the Seventh International Conference on Networking and Services, May 22, 2011, pp. 32 to 37.

[6]   R. C. Carrano, Magalha, x, L. C. S. es, D. C. M. Saade, and C. V. N. Albuquerque, "IEEE 802.11s Multihop MAC: A Tutorial," *IEEE Communications Surveys and Tutorials*, vol. 13, no. 1, pp. 52-67, 2011.

[7]   R. Garroppo, S. Giordano, D. Iacono, and L. Tavanti, "Notes on Implementing a IEEE 802.11s Mesh Point," in *Wireless Systems and Mobility in Next Generation Internet*, vol. 5122, *Lecture Notes in Computer Science*, L. Cerdà-Alabern, Ed.: Springer Berlin Heidelberg, 2008, pp. 60-72.

[8]   G. R. Hiertz, S. Max, Z. Rui, D. Denteneer, and L. Berlemann, "Principles of IEEE 802.11s," presented at the 16th International Conference on Computer Communications and Networks, August 13-16, 2007, pp. 1002-1007.

[9]   D. Skordoulis, N. Qiang, C. Hsiao-Hwa, A. P. Stephens, L. Changwen, and A. Jamalipour, "IEEE 802.11n MAC frame aggregation mechanisms for next-generation high-throughput WLANs," *IEEE Wireless Communications*, vol. 15, no. 1, pp. 40-47, 2008.

[10] A. Ksentini, A. Guéroui, and M. Naimi, "Adaptive transmission opportunity with admission control for IEEE 802.11e networks," in *the 8th ACM international symposium on Modeling, analysis and simulation of wireless and mobile systems*. Montrial, Quebec, Canada: ACM, 2005, pp. 234-241.

[11] S. Ould Cheikh and A. Gueroui, "Multi-hop bandwidth reservation in WMN-based IEEE 802.11s (MBRWMN)," presented at the International Conference on Communications and Information Technology (ICCIT), June 26-28, 2012, pp. 211-215.

[12] D. B. Rawat, D. C. Popescu, and S. Min, "Performance enhancement of EDCA access mechanism of IEEE 802.11e wireless LAN," presented at the IEEE Radio and Wireless Symposium, January 22-24, 2008, pp. 507-510.

[13] H. Jie and M. Devetsikiotis, "Performance analysis of IEEE 802.11e EDCA by a unified model," presented at the IEEE Global Telecommunications Conference, November 29 - December 3, 2004, pp. 754-759.

[14] J. Hoblos, "Fairness enhancement in IEEE 802.11s multi-hop wireless mesh networks," presented at the 13th IEEE International Conference on Communication Technology (ICCT), Sept 25-28, 2011, pp. 647-651.

[15] K. Duffy, D. J. Leith, T. Li, and D. Malone, "Modeling 802.11 mesh networks," *IEEE Communications Letters*, vol. 10, no. 8, pp. 635-637, 2006.

[16] V. Mancuso, O. Gurewitz, A. Khattab, and E. W. Knightly, "Elastic Rate Limiting for Spatially Biased Wireless Mesh Networks," presented at the IEEE INFOCOM, March 14-19, 2010, pp. 1-9.

[17] J. Trivic, "Simulation of user plane in IEEE 802.11s mesh networks," presented at the 19th Telecommunications Forum (TELFOR), November 22-24, 2011, pp. 1624-1627.

[18] C. Sunghyun, J. del Prado, N. Sai Shankar, and S. Mangold, "IEEE 802.11 e contention-based channel access (EDCF) performance evaluation," presented at IEEE International Conference on Communications, Anchorage, AK, May 11-15, 2003, pp. 1151-1156.

[19] G. R. Hiertz, S. Max, T. Junge, D. Denteneert, and L. Berlemann, "IEEE 802.11s - Mesh Deterministic Access," presented at the 14th European Wireless Conference, June 22-25, 2008, pp. 1-8.

[20] C. Cicconetti, L. Lenzini, and E. Mingozzi, "Scheduling and Dynamic Relocation for IEEE 802.11s Mesh Deterministic Access," presented at the 5th annual IEEE Communications Society Conference on Sensor, Mesh and Ad Hoc Communications and Networks, June 16-20, 2008, pp. 19-27.

[21] T. Li, Ni,Qiang, Xiao,Yang, "Investigation of the block ACK scheme in wireless ad hoc networks," *Wireless Communications and Mobile Computing*, vol. 6, no. 6, pp. 877-888, 2006.

[22] L. Seungbeom and P. Sin-Chong, "Rotating Priority Queue based Scheduling Algorithm for IEEE 802.11n WLAN," *The 9th International Conference on Advanced Communication Technology*, vol. 3, no. 1702-1706, 2007.

[23] T. Hiatt and A. Prodan, "Investigating channel bonding and TXOP in 802.11n wireless networks," presented at the IEEE TENCON, Nov 21-24, 2011, pp. 435-439.

[24] O. Hoffmann and R. Kays, "Efficiency of frame aggregation in wireless multimedia networks based on IEEE 802.11n," presented at the 14th IEEE International Symposium on Consumer Electronics (ISCE), 7-10 June, 2010, pp. 1-5.

[25] B. Ginzburg and A. Kesselman, "Performance analysis of A-MPDU and A-MSDU aggregation in IEEE 802.11n," presented at the IEEE Sarnoff Symposium, April 30 - May 2, 2007, pp. 1-5.

[26] A. Saif, M. Othman, S. Subramaniam, and N. Hamid, "An Enhanced A-MSDU Frame Aggregation Scheme for 802.11n Wireless Networks," *Wireless Personal Communications*, vol. 66, no. 4, pp. 683-706, 2012.

[27] O. Eng Hwee, J. Kneckt, O. Alanen, C. Zheng, T. Huovinen, and T. Nihtila, "IEEE 802.11ac: Enhancements for very high throughput WLANs," presented at the 22nd IEEE International Symposium on Personal Indoor and Mobile Radio Communications (PIMRC), September 11-14, 2011, pp. 849-853.

[28] K. Yaw-Wen, L. Tsern-Huei, H. Yu-Wen, and H. Jing-Rong, "Design and evaluation of a high throughput MAC with QoS guarantee for wireless LANs," presented at IEEE 9th Malaysia International Conference on Communications (MICC), December 15-17, 2009, pp. 869-873.

[29] M. Alicherry, R. Bhatia, and L. Li Erran, "Joint Channel Assignment and Routing for Throughput Optimization in Multiradio Wireless Mesh Networks," *Journal on Selected Areas in Communications*, vol. 24, no. 11, pp. 1960-1971, 2006.

[30] E. Khorov, A. Kiryanov, A. Lyakhov, and A. Safonov, "Analytical study of link management in IEEE 802.11s mesh networks," presented at the International Symposium on Wireless Communication Systems (ISWCS), August 28-31, 2012, pp. 786-790.

[31] Z. Wenxuan, W. Jing, and K. Guixia, "A novel High Throughput Long Training Field sequence design for next-generation WLAN," presented at Wireless Telecommunications Symposium (WTS), April 13-15, 2011, pp. 1-5.

[32] C.-Y. Wang and H.-Y. Wei, "IEEE 802.11n MAC Enhancement and Performance Evaluation," *Mobile Networks and Applications*, vol. 14, no. 6, pp. 760-771, 2009.

[33] K. Seongkwan, K. Youngsoo, C. Sunghyun, J. Kyunghun, and C. Jin-Bong, "A high-throughput MAC strategy for next-generation WLANs," presented at the 6th IEEE International Symposium on World of Wireless Mobile and Multimedia Networks, June 13-16, 2005, pp. 278-285.

[34] X. Wang and A. O. Lim, "IEEE 802.11s wireless mesh networks: Framework and challenges," *Ad Hoc Networks*, vol. 6, no. 6, pp. 970-984, 2008.

[35] T. Selvam and S. Srikanth, "A frame aggregation scheduler for IEEE 802.11n," presented at the National Conference on Communications (NCC), January 29-31, 2010, pp. 1-5.

[36] J. Camp and E. Knightly, "The IEEE 802.11s Extended Service Set Mesh Networking Standard," *IEEE Communications Magazine*, vol. 46, no. 8, pp. 120-126, 2008.

[37] I. Tinnirello and G. Bianchi, "Rethinking the IEEE 802.11e EDCA Performance Modeling Methodology," *IEEE/ACM Transactions on Networking*, vol. 18, no. 2, pp. 540-553, 2010.

[38] L. Bih-Hwang, C. Hung-Chi, and W. Huai-Kuei, "Study on multi-channel deterministic access for wireless mesh LAN," presented at the IEEE International Conference on Cyber Technology in Automation, Control, and Intelligent Systems (CYBER), May 27-31, 2012, pp. 39-42.

[39] R. G. Sanchez, H. Xiaojing, and C. Kwan-Wu, "Viability of concurrent transmission and reception for UWB radios over multipath channels," presented at the International Symposium on Communications and Information Technologies, October 17-19, 2007, pp. 1241-1246.

[40] Z. Yingnan, Z. Wenjun, L. Hang, G. Yang, and S. Mathur, "Supporting Video Streaming Services in Infrastructure Wireless Mesh Networks: Architecture and Protocols," presented at the IEEE International Conference on Communications, May 19-23, 2008, pp. 1850-1855.

[41] P. Dely, A. Kassler, N. Bayer, and D. Sivchenko, "An Experimental Comparison of Burst Packet Transmission Schemes in IEEE 802.11-Based Wireless Mesh Networks," presented at the IEEE Global Telecommunications Conference, December 6-10, 2010, pp. 1-5.

[42] N. S. Nandiraju, D. S. Nandiraju, D. Cavalcanti, and D. P. Agrawal, "A novel queue management mechanism for improving performance of multihop flows in IEEE 802.11s based mesh networks," presented at the 25th IEEE International Performance, Computing, and Communications Conference, April 10-12, 2006, pp. 7 pp.-168.

[43] H. Ming-Xin and K. Geng-Sheng, "Delay and throughput Analysis of IEEE 802.11s Networks," presented at the IEEE International Conference on Communications Workshops, May 19-23, 2008, pp. 73-78.

[44] M. Singh, S.-G. Lee, W. Tan, and J. Lam, "Throughput Analysis of Wireless Mesh Network Test-Bed," in *Convergence and Hybrid Information Technology*, vol. 206, *Communications in Computer and Info Science*, G. Lee, D. Howard, and D. Ślęzak, Eds.: Springer Berlin Heidelberg, 2011, pp. 54-61.

[45] B. Staehle, M. Bahr, F. Desheng, and D. Staehle, "Intra-Mesh Congestion Control for IEEE 802.11s Wireless Mesh Networks," presented at the 21st International Conference on Computer Communications and Networks (ICCCN), July 30 - August 2, 2012, pp. 1-7.

[46] L. Zheng, Y. Min, D. Heng, and D. Jufeng, "Concurrent Transmission Scheduling for Multi-Hop Multicast in Wireless Mesh Networks," presented at the 4th International Conference on Wireless Communications, Networking and Mobile Computing, October 12-14, 2008, pp. 1-4.

[47] M. Ozdemir, G. Daqing, A. B. McDonald, and Z. Jinyun, "Enhancing MAC Performance with a Reverse Direction Protocol for High-Capacity Wireless LANs," presented at the 64th IEEE Vehicular Technology Conference, September 25-28, 2006, pp. 1-5.

[48] D. Akhmetov, "802.11N: Performance Results of Reverse Direction Data Flow," presented at the IEEE 17th International Symposium on Personal, Indoor and Mobile Radio Communications, September 11-14, 2006, pp. 1-3.

8

Secure Data in Wireless Sensor Network By Using DES

Jagbir Dhillon , Krishna Prasad , Rajesh Kumar, Ashok Gill

jagbirdhillon@yahoo.co.in

ABSTRACT

The main goal of sensor networks is to provide precise information about a sensing field for an extended period of time. The emergence of sensor networks is one of the dominant technical trend has posed challenges to explorers. As sensor networks may interact with sensitive data and operate in hostile insecure environment, it is clear that these security concerns can be addressed from the basic system design. These networks are likely to be composed of hundreds, and potentially thousands of small sensor nodes, functioning independently .The challenges in sensor networks are diverse, we concentrate on security of Wireless Sensor Network in this paper [1,2,3,4,5]. Some of the security goals are proposed by our self for Wireless Sensor Network and use of sensor networks for many applications. We also propose some techniques against these threats in Wireless Sensor Network. So, in this paper we have implemented Encryption Algorithm like - DES to provide sufficient levels of security for protecting the confidentiality of the data in the WSN network. This paper also analyzes the performance of DES algorithm against Attacks in WSN Network [3,5].

KEYWORDS: WSN, Sensor node, Gateway, Security, DES.

1. INTRODUCTION:

Wireless sensor networks are very popular because of the fact that they are offer low cost solutions to a variety of real-world challenges. Due to low cost one can deploy large sensor arrays in both military and civilian areas. But sensor networks also face severe resource constraints due to their lack of data storage and power and hence implementation of traditional computer security techniques in a wireless sensor network become difficult [8, 9, 11]. The unreliable and unattended communication channel makes the security defenses even difficult. Often wireless sensors have the processing characteristics of machines that are very old, whereas the industry wants to reduce the cost of wireless sensors while maintaining similar computing power. Researchers have begun to maximize the processing capabilities and energy efficiency of wireless sensor nodes while also securing them against attacks. All aspects regarding wireless sensor network are being analyzed including secure and optimal routing, data aggregation, cluster formation, and so on. Apart from these traditional security issues, many sensor network techniques assumed that all nodes are trustworthy. Some of the researchers began to concentrate on building a sensor network model to solve the desired problems by using cryptographic schemes. Due to unattended feature of wireless sensor networks, we see that physical attacks to sensors play an important role in the operation of wireless sensor networks. Thus, we include a detailed discussion of the physical attacks and their corresponding defenses, topics typically ignored in most of the current research on sensor security. We classify wireless sensor network security in four categories: the obstacles to sensor network security, the need of a secure wireless sensor network, attacks, and defensive measures. We also give a brief introduction of related security techniques and summarize the obstacles for the sensor network security [13, 14, 15]. The security requirements of a wireless sensor network are listed as below:

1.1. Obstacles of Sensor Security

A wireless sensor network is a network which has many constraints compared to a traditional computer network and these constraints make it difficult to use the existing security techniques to wireless sensor networks. However, it is necessary to develop useful security mechanisms by getting ideas from security techniques like (DES, AES).

2. WSN ARCHITECTURE

 In a typical WSN we see following network components –

[A]. Sensor motes– Routers mounted in the process must be capable of routing packets on behalf of other devices. They control the process and process equipments. A router is a field device which do not have process sensor or control equipment and does not interface with the process.

[B]. Gateway or Access points – A Gateway provides communication between Host application and field devices.

[C].Network manager –To maintain the scheduled communication between devices and configuration of the network for which Network Manager is essential. Also manages the routing tables and monitoring the condition of the network.

[D].Security manager – The Security Manager manages the storage and management of the Cryptographic keys like DES and AES. [5, 18, 19].

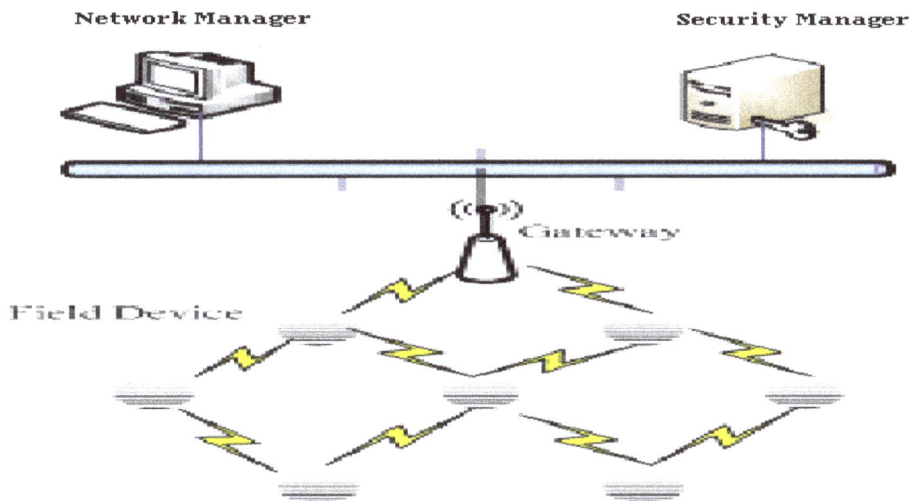

Figure 1 WSN Architecture

3. WSN SECURITY ANALYSIS

Wireless Sensor Network are vulnerable to variety of attacks due to their simplicity. To secure wireless sensor network the network should support all security issues: confidentiality, integrity, authenticity and availability. Attackers may deploy a few malicious nodes with similar hardware configuration as the genuine nodes that might plan to attack the system. Attackers may approach malicious nodes by purchasing these separately. Also, these nodes have better communications links for coordinating these attack. Sensor nodes if attacked , one can extract all key materials, data, and code . So tamper resistance might be a defense for some networks, we do not see it as a general purpose solution. On the other hand sensor nodes are intended to be very inexpensive [21, 23]

3.1 Overview

The Data Encryption Standard is the Federal Information Processing Standard, which gives the data encryption algorithm. As per ANSI standard X3.92, data encryption algorithm is an improvement of the algorithm Lucifer developed by IBM in 1970s. IBM and the National Security Agency developed the algorithm. The DES has studied in depth and is the most widely used symmetric algorithm in the world. [15, 5, 19]. DES has a 64-bit block size and uses a 56-bit key during execution. DES is a symmetric cryptosystem, specifically a 16-round Feistel cipher.

As we know that WSN is mostly used for the application of collection of information from the surrounding environment, so it is necessary to protect the sensitive data from unauthorized parties. WSNs are susceptible to security attacks due to the broadcast radio transmission. It is very clear to all of us that some of the sensor nodes are physically captured or destroyed by the enemies. The main role of sensor network in security for various applications depends on secure routing. The resources of sensor network show many challenges for its security. As sensor nodes are very limited for computing power and it is very difficult to provide security in WSN using public-key cryptography because it is very costly. So security solutions for various applications are based on symmetric key cryptography. In this paper we have used DES for the security purpose of wireless sensor network. [22, 23]

3.2 OVERVIEW OF SECURITY ISSUES

3.2.1. Attack and attacker

An attack can be defined as an attempt to gain unauthorized access to a service, a resource or information, or the attempt to compromise integrity, availability, or confidentiality of a system. As we know that all the attackers are the originator of an attack. Main weaknesses in a system are security design, configuration, implementation or limitations which can be exploited by the attackers are known as vulnerability or flaw. Any event or attacker which impacts a system through the security breach is called threat. So the attacker will exploit a particular vulnerability and causing harm to a system asset is called as risk.

3.2.2. Security requirements

A sensor network is a type of Ad hoc network. The security requirements of a wireless sensor network can be classified as follows [26, 27, 29, 30]

Authentication: As we know that WSN communicates sensitive data that helps in making the important decision. The receiver needs to ensure that the data used originates from the desired source. During exchange of control information in the WSN, authentication is important.

Integrity: Data transit may be changed by the attackers or intruders. Data loss can also occur due to the harsh communication environment. Data integrity ensures that no information is changed in transit due to malicious node or by accident.

Data Confidentiality: Some of the applications need to rely on confidentiality like surveillance of information, industrial secrets and key distribution etc. Encryption standards are used to keep data confidentiality.

Data Freshness: We need to ensure freshness of all messages even if confidentiality and data integrity are assured. Data freshness means that the data is recent and no old messages are replayed. A time stamp can be added to the packet to ensure that no old messages are replayed.

Availability: Due to excess computation or communication, battery power of sensor nodes may run out and become unavailable because attackers may jam communication. The requirement of security is very important in maintaining the availability of the network.

Self-Organization: As we know that in WSN every sensor node should be independent and flexible to be self-organizing and self-healing according to different environment conditions. Therefore no fixed infrastructure is available for WSN network due to random deployment of nodes & distributed sensor networks must self-organize to support optimal routing in WSN.

Time Synchronization: Most sensor network applications require time synchronization. An individual sensors radio may be turned off periodically to conserve power.

Secure Localization: The sensor network always needs information of location accurately. Therefore by reporting false signal strengths and replaying signals, an attacker can easily manipulate non-secured location information in WSN. [21, 28]

3.3 In Depth

By using a 64-bit key which gives a 64-bit block of plaintext as input and gives an output in the form of a 64-bit block of cipher text. DES operates on blocks of equal size by using permutations and substitutions in the algorithm. Because DES has 64-bit rounds, hence the main algorithm is repeated 16 times to produce the cipher text. However it has been noted that the number of rounds are always exponentially proportional to the time required. Security of the algorithm increases exponentially when the number of rounds increases which means security is maximized automatically when the number of rounds are more. [3, 4, 5].

64 bit plain text
64 bit plain text
64 bit plain text
DES
DES
DES
64 bit cipher text
64 bit cipher text
64 bit cipher text

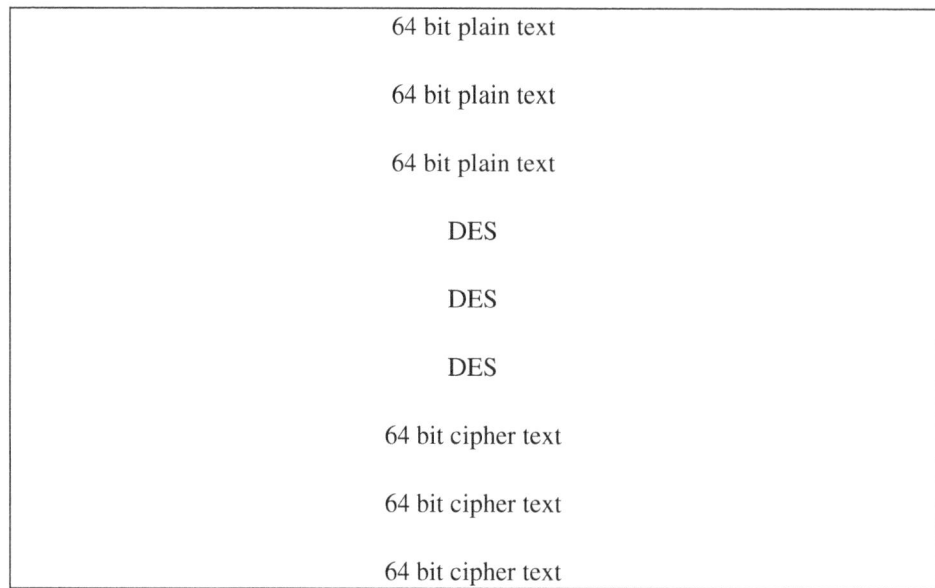

Figure2. Conceptual working of DES

3.4 Key Scheduling

DES has 64-bits long input key but the in real practice key used by DES is of 56-bits length. A parity bit is LSB in each byte set in such a way that there should be odd number of 1s in every byte. As these parity bits are ignored, hence the seven most significant bits are used for each byte giving us 56-bits key results. Firstly the 64-bit key is passed through a Permuted Choice 1(PC1). The table is given below. All descriptions of bit numbers, 1 is the left-most bit and n is the rightmost bit.

PC -1: Permuted Choice 1						
Bit	0	1	2	3	4	5	6
1	57	49	41	33	25	17	9
8	1	58	50	42	34	26	18
15	10	2	59	51	43	35	27
22	19	11	3	60	52	44	36
29	63	55	47	39	31	23	15
36	7	62	54	46	38	30	22
43	14	6	61	53	45	37	29
50	21	13	5	28	20	12	4

Table 1: Permuted Choice 1

For example, PC-1 table can be used to see how bit 30 of the original 64-bit key changes to a new 56-bit key. As number 30 in the table, belongs to the column no.5 and the row no. 36. Now adding values of the row and column to know the new position of the bit. For bit 30, 36+5=41, so bit 30 becomes bit 41 of the new 56-bit key. Therefore bit 8,16,24,32,40,48,56 and 64 of the original key are not present in the table. They are the unused parity bits which are discarded if the final 56-bit key is created [17, 19, 5].

1	2	3	4	5	6	7	8	9	10	11	12	13	14	15	16
17	18	19	20	21	22	23	24	25	26	27	28	29	30	31	32
33	34	35	36	37	38	39	40	41	42	43	44	45	46	47	48
49	50	51	52	53	54	55	56	57	58	59	60	61	62	63	64

Figure3. Discarding of every 8th bit of Original Key (Shaded Bit Position are Discarded)

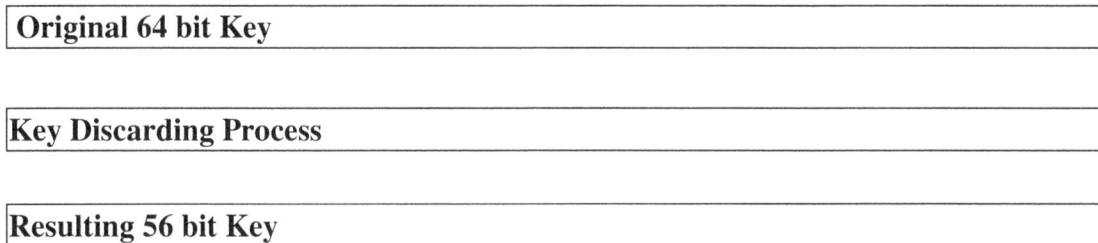

Original 64 bit Key

Key Discarding Process

Resulting 56 bit Key

Figure 4. Key Discarding Process

Now that we have the 56-bit key, the next step is to use this key to generate 16 48-bit sub keys, called K [1] – K [16], which is used in the 16 rounds of DES for encryption and decryption. The procedure for generating the sub keys – known as key scheduling – is fairly simple:

Set the round number R to 1.

Split up the present 56-bit key, K, into two 28-bit blocks, L and R.

Rotate L left by the number of bits given in the table below, and repeat the same procedure for rotating R left by the same number.

Join L and R together to get the new K.

Apply Permuted Choice 2 (PC-2) to K to get the final K[R], where R is the round number we are on.

Give increment of 1 to R and repeat its procedure until we get 16 sub keys, i.e., up to K [1] to K [16].

Here are the tables involved in these operations:

Sub Key Rotation Table:

Round Number	1	2	3	4	5	6	7	8	9	10	11	12	13	14	15	16
Number of Bits to Rotate	1	1	2	2	2	2	2	2	1	2	2	2	2	2	2	1

Table 2: Sub Key Rotation Table

P C 2 : Permuted Choice 2					
Bit	0	1	2	3	4	5
1	14	17	11	24	1	5
7	3	28	15	6	21	10
13	23	19	12	4	26	8
19	16	7	27	20	13	2
25	41	52	31	37	47	55
31	30	40	51	45	33	48
37	44	49	39	56	34	53
43	46	42	50	36	29	32

Table 3: P C 2 : Permuted Choice 2

3.5 Plaintext Preparation

After the key scheduling has been done, the next step is to encrypt. The plaintext is passed through a permutation called as the Initial Permutation (IP). This table also has the Inverse Initial Permutation, or IP^ (-1) known as the Final Permutation. These are given below.

IP: Initial Permutation							
Bit	0	1	2	3	4	5	6	7
1	58	50	42	34	26	18	10	2
9	60	52	44	36	28	20	12	4
17	62	54	46	38	30	22	14	6
25	64	56	48	40	32	24	16	8
33	57	49	41	33	25	17	9	1
41	59	51	43	35	27	19	11	3
49	61	53	45	37	29	21	13	5
57	63	55	47	39	31	23	15	7

Table 4: IP(Initial Permutation)

IP^(-1):Inverse Initial Permutation							
Bit	0	1	2	3	4	5	6	7
1	40	8	48	16	56	24	64	32
9	39	7	47	15	55	23	63	31

17	38	6	46	14	54	22	62	30
25	37	5	45	13	53	21	61	29
33	36	4	44	12	52	20	60	28
41	35	3	43	11	51	19	59	27
49	34	2	42	10	50	18	58	26
57	33	1	41	9	49	17	57	25

Table 5: IP^(-1) Inverse Initial Permutation

Tables mentioned above are used for the key scheduling. By looking at the tables it is very clear why one permutation is called the inverse of the other. For example, how bits 32 are transformed under IP. Bit 32 available at the intersection of the column 4 and row 25 as shown above, apply IP^ (-1). In IP^ (-1), bit 29 which is located at the intersection of the column no. 7 and the row no. 25. So this bit becomes bit 32 (25+7) after the permutation is done which is the same bit position that we have started before permutation. So inverse of IP^ (-1) is IP. It does the exact opposite of IP. We will end up with the original block.

3.6 DES Core Function

Once key scheduling and plaintext preparation are over, then actual encryption or decryption is performed by using DES main algorithm. The 64-bit block of input data is first split into two halves, L and R. L is the left-most 32 bits, and R is the right-most 32 bits.

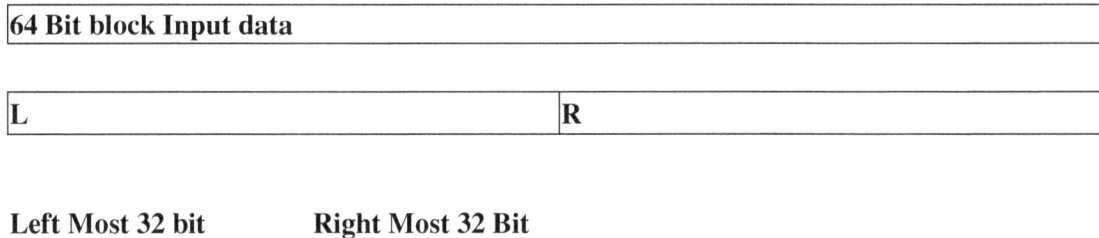

64 Bit block Input data

L	R

Left Most 32 bit   **Right Most 32 Bit**

Figure 5. Splitting 64-bit block input data

The following is repeated 16 times for making up the 16 rounds of standard. We call the 16 sets of halves L[0] to L[15] and R[0] to R[15].

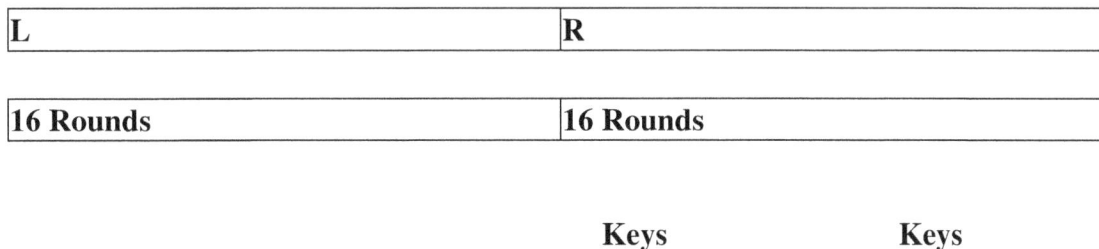

L	R

16 Rounds	16 Rounds

Keys   **Keys**

Figure 6. 16 Rounds

The following process is repeated 16 times for making up the 16 rounds of standard DES. We call the 16 sets of halves L [0] – L [15] and R [0] – R [15].

This expands the number R [I-1] from 32 to 48 bits to prepare for the next step.

E-Bit Selection Table						
Bit	0	1	2	3	4	5
1	32	1	2	3	4	5
7	4	5	6	7	8	9
13	8	9	10	11	12	13
19	12	13	14	15	16	17
25	16	17	18	19	20	21
31	20	21	22	23	24	25
37	24	25	26	27	28	29
43	28	29	30	31	32	1

Table 6: E-Bit Selection Table

The 48-bit R [I-1] is XORed with K [I] and stored in a temporary buffer so that R [I-1] is not modified.

The result from the last step is now split up into 8 parts of 6 bits each. The left-most 6 bits are denoted as B [1], and the right-most 6 bits as B [8]. These blocks from the index into the S-boxes are to be used in the next step. The Substitution boxes(S-boxes) are a set of eight 2-dimensional arrays, with 4 rows and 16 columns. The numbers in the boxes are always 4 bits in length and their values from 0 to 15. The S-boxes are numbered S [1] to S [8].

For B [1], first and last bits of the 6-bit block are used as an index into the row number of S [1], with range from 0 to 3, and middle four bits are used as an index into the column number, with its range from 0 to 15. The number from this position in the S-box is taken and stored which is repeated with B [2] and S [8]. At this point, now we have eight 4-bit numbers, which give a 32-bit result.

Now applying permutation in the result of previous stage as all details given in next step

This number is now XORed with L [I-1], and moved into R [I]. R [I-1] is moved into L [I].

At this point we have a new L [T] and R [I]. Here, increment is given to I and same core function is repeated until we get I = 17, So that 16 rounds have been executed and all the keys K[I] to K[16] are used.

When L [16] and R [16] have been obtained, they are joined back together in the same fashion they were split apart (L [16] is the left-hand half, R [16] is the right-half hand), then the two halves are swapped, R [16] becomes the left-most 32 bits and L [16] becomes the right-most 32 bits of the pre-output block and the resultant 64- bit number is called the pre-output.

P- Permutation			
Bit	0	1	2	3
1	16	7	20	21
5	29	12	28	17
9	1	15	23	26
13	5	18	31	10
17	2	8	24	14
21	32	27	3	9
25	19	13	30	6
29	22	11	4	25

S-Box 1:Substitution Box 1															
Bit	0	1	2	3	4	5	6	7	8	9	10	11	12	13	14	15
0	14	4	13	1	2	15	11	8	3	10	6	12	5	9	0	7
1	0	15	7	4	14	2	13	1	10	6	12	11	9	5	3	8
2	4	1	14	8	13	6	2	11	15	12	9	7	3	10	5	0
3	15	12	8	2	4	9	1	7	5	11	3	14	10	0	6	13

S-Box 2:Substitution Box 2															
Bit	0	1	2	3	4	5	6	7	8	9	10	11	12	13	14	15
0	15	1	8	14	6	11	3	4	9	7	2	13	12	0	5	10
1	3	13	4	7	15	2	8	14	12	0	1	10	6	9	11	5
2	0	14	7	11	10	4	13	1	5	8	12	6	9	3	2	15
3	13	8	10	1	3	15	4	2	11	6	7	12	0	5	1	4

S-Box 3:Substitution Box 3															
Bit	0	1	2	3	4	5	6	7	8	9	10	11	12	13	14	15
0	10	0	9	14	6	3	15	5	1	13	12	7	11	4	2	8
1	13	7	0	9	3	4	6	10	2	8	5	14	12	11	15	1
2	13	6	4	9	8	15	3	0	11	1	2	12	5	10	14	7
3	1	10	13	0	6	9	8	7	4	15	14	3	11	15	2	12

S-Box 4:Substitution Box 4															
Bit	0	1	2	3	4	5	6	7	8	9	10	11	12	13	14	15
0	7	13	14	3	0	6	9	10	1	2	8	5	11	12	4	15
1	13	8	11	5	6	15	0	3	4	7	2	12	1	10	14	9
2	10	6	9	0	12	11	7	13	15	1	3	14	5	2	8	4
3	3	15	0	6	10	1	13	8	9	4	5	11	12	7	2	14

S-Box 5:Substitution Box 5

Bit	0	1	2	3	4	5	6	7	8	9	10	11	12	13	14	15
0	2	12	4	1	7	10	11	6	8	5	3	15	13	0	14	9
1	14	11	2	12	4	7	13	1	5	0	15	10	3	9	8	6
2	4	2	1	11	10	13	7	8	15	9	12	5	6	3	0	14
3	11	8	12	7	1	14	2	13	6	15	0	9	10	4	5	3

S-Box 6:Substitution Box 6

Bit	0	1	2	3	4	5	6	7	8	9	10	11	12	13	14	15
0	12	1	10	15	9	2	6	8	0	3	13	4	14	7	5	11
1	10	15	4	2	7	12	9	5	6	1	13	14	0	11	3	8
2	9	14	15	5	2	8	12	3	7	0	4	10	1	13	11	6
3	4	3	2	12	9	5	15	10	11	14	1	7	6	0	8	13

S-Box 7:Substitution Box 7

Bit	0	1	2	3	4	5	6	7	8	9	10	11	12	13	14	15
0	4	11	2	14	15	0	8	13	3	12	9	7	5	10	6	1
1	13	0	11	7	4	9	1	10	14	3	5	12	2	15	8	6
2	1	4	11	13	12	3	7	14	10	15	6	8	0	5	9	2
3	6	11	13	8	1	4	10	7	9	5	0	15	14	2	3	12

S-Box 8:Substitution Box 8

Bit	0	1	2	3	4	5	6	7	8	9	10	11	12	13	14	15
0	13	2	8	4	6	15	11	1	10	9	3	14	5	0	12	7
1	1	15	13	8	10	3	7	4	12	5	6	11	0	14	9	2
2	7	11	4	1	9	12	14	2	0	6	10	13	15	3	5	8
3	2	1	14	7	4	10	8	13	15	12	9	0	3	5	6	11

Table 7: S-Boxes

3.7 How to use the S-boxes

The aim of the example is to explain the working of the S-boxes. Suppose we have the following 48-bit binary number:

011101000101110101000111101000011100101101011101

In order to pass this through steps 3 and 4 of the Core Function as outlined above, the number is split up into 8 6-bit blocks, labeled B[1] to B[8] from left to right:

011101 000101 110101 000111 101000 011100 101101 011101

Now, eight numbers are extracted from the S- boxes – one from each box:

B[1] = S[1](01,1110) = S[1][1][14]   = 3   = 0011

B[2] = S[2](01,0010) = S[2][1][2]   = 4   = 0100

B[3] = S[3](11,1010) = S[3][3][10]   = 14   = 1110

B[4] = S[4](01,0011) = S[4][1][3]   = 5   =0101

B[5] = S[5](10,0100) = S[5][2][4]   = 10   = 1010

B[6] = S[6](00,1110) = S[6][0][14]   = 5   = 0101

B[7] = S[7](11,0110) = S[7][3][6]   = 10   = 1010

B[8] = S[8](01,1110) = S[8][1][14]   = 9   = 1001

Row index of B[n] is first and last bit and column index of S[n] are the middle four bits.

The results are now joined together to form a 32 – bit number which serves as the input to stage 5 of the Core Function (the P Permutation):

00110100111001011010010110101001

4. Conclusion:

Data encryption is utilized in various applications and environments. The specific utilization of encryption and the implementation of the DES will be based on many factors particularly to the WSN and its associated components like sensor node, Gateway, Secure routing in WSN etc. Cryptography is used to protect data while it is communicating between two points or while it is stored in a medium vulnerable to physical theft. So data encryption standard is very useful for secure routing in wireless sensor network. For security point of view, we can also use Advanced encryption standard (AES) to secure routing in wireless sensor network. As we know that none of the protocol is designed to keep security in mind, so there should be a proper implementation and design in routing protocol for security purpose in wireless sensor network. I will concentrate on DES for security purpose in my research topic of WSN.

References:

[1]. International Journal of Technology and Applied Science, Vol. 2, pp. 5-11, 2011. ISSN: 2230-9004 © 2011 IJTAS 5 Repairing the Gaps in Connectivity of Wireless Sensor Network & WiMAX Using Robots: Jagbir Dhillon, Krishna Parsad, Rajesh Kumar.

[2].International Journal of VLSI and Signal Processing Applications, Vol. 1, Issue 1 (48- 61),ISSN 2231-3133 48A Survey on the state of the art of secure and optimal routing issues in wireless sensor networks. *Jagbir Dhillon, Prasad K.P, Krishan Kumar* jagbirdhillon@yahoo.co.in

[3]. Distributed Recovery from Network Partitioning in Movable Sensor/Actor Networks via Controlled Mobility, Kemal Akkaya, Member, IEEE, Fatih Senel, Aravind Thimmapuram, and Suleyman Uludag, Member, IEEE, IEEE transactions on computers, vol. 59, no. 2, february 2010.

[4]. IEEE journal on selected areas in communications, vol. 28, no. 5, june 2010 Cross Layer QoS-Aware Communication for Ultra Wide Band Wireless Multimedia Sensor Networks Tommaso Melodia, Member, IEEE, and Ian F. Akyildiz, Fellow, IEEE.

[5]. IEEE journal on selected areas in communications, vol. 28, no. 7, september 2010. Handling Inelastic Traffic in Wireless Sensor Networks Jiong Jin, Student Member, IEEE, Avinash Sridharan, Bhaskar Krishnamachari, Member, IEEE and Marimuthu Palaniswami, Senior Member, IEEE.

[6]. Constrained Relay Node Placement in Wireless Sensor Networks: Formulation and Approximations, Satyajayant Misra, Member, IEEE, Seung Don Hong, Guoliang (Larry) Xue, Senior Member, IEEE, and Jian Tang, Member, IEEE; IEEE/ACM transactions on networking, vol. 18, no. 2, April 2010.

[7]. Deploying Sensor Networks With Guaranteed Fault Tolerance; Jonathan L. Bredin, Erik D. Demaine, Mohammad Taghi Hajiaghayi, and Daniela Rus; IEEE/ACM transactions on networking, vol. 18, no. 1, february 2010.

[8]. A Distributed Node Localization Scheme for Wireless Sensor Networks; Qinqin Shi, Hong Huo, Tao Fang, Deren Li; Published online: 26 March 2009 © Springer Science+Business Media, LLC. 2009.

[9]. High Reliable In-Network Data Verification in Wireless Sensor Networks; Dong-Wook Lee, Jai-Hoon Kim; Published online: 29 May 2009 © Springer Science+Business Media, LLC. 2009.

[10]. Secure and Efficient Localization Scheme in Ultra-Wideband Sensor Networks; Daojing He, Lin Cui, Hejiao Huang, Maode Ma; Published online: 4 November 2008 © Springer Science+Business Media, LLC. 2008.

[11]. TMH Book of Cryptography & System Security.

[12]. Phalguni Gupta Internet & Protection Security Book.

[13] Adrian Perrig, John Stankovic, David Wagner, "Security in Wireless Sensor Networks" Communications of the ACM, Page53-57.

[14] Al-Sakib Khan Pathan, Hyung-Woo Lee, Choong Seon Hong, "Security in Wireless Sensor Networks: Issues and Challenges", International conference on Advanced Computing Technologies, Page1043-1045.

[15] A. S. Wander, N. Gura, H. Eberle, V. Gupta, and S. C. Shantz, "Energy analysis of public-key cryptography for wireless sensor networks," in Third IEEE International Conference on Pervasive Computing and Communications (PERCOM'05). IEEE Computer Society Press, pp. 324-328.

[16] D. C. Schleher, Electronic Warfare in the Information Age. Artech.

[17] D. Djenouri, L. Khelladi, and N. Badache, "A Survey of Security Issues in Mobile ad hoc and Sensor Networks," IEEE Commun. Surveys Tutorials, vol. 7, pp. 2–28.

[18] D.Ganesan, R.Govindan, S.Shenker, and D.Estrin, "Highly resilient, energy efficient multipath routing in wireless sensor networks," Mobile Computing and Communications Review (MC2R), vol. 1, no. 2.

[19] F. Nait-Abdesselam, B. Bensaou, T. Taleb, "Detecting and avoiding wormhole attacks in wireless Ad hoc networks," IEEE Communication Magazine, Vol.46, Issue 4, pp. 127-133, April 2008.

[20] H. Yang, H. Luo, F. Ye, S. Lu, and L. Zhang, "Security in mobile ad hoc networks: challenges and solutions," IEEE Wireless Communications, vol. 11, no. 1, pp. 38- 47.

[21] Ian F. Akykildiz, Weilian Su, Yogesh Sankarasubramaniam, and Erdal Cayirci, "A Survey on Sensor Networks", IEEE Communication Magazine.

[22] John Paul Walters, Zhengqiang Liang, Weisong Shi, Vipin Chaudhary, "Wireless Sensor Network Security: A Survey", Security in Distributed, Grid and Pervasive Computing Yang Xiao (Eds), Page3-5, 10-15.

[23] Mohit Saxena, "Security In Wireless Sensor Networks - A Layer Based Classification", Cerias Tech Report 2007-04. [24] S. Khan, K-k. Loo, T. Naeem, M.A. Khan, "Denial of service attacks and challenges in broadband wireless network," International Journal of Computer Science and Network Security, Vol. 8, No. 7, pp.1-6.

[25] Wang, B-T. and Schulzrinne, H., "An IP trace back mechanism for reflective DoS attacks", Canadian M. Conference on Electrical and Computer Engineering, Volume 2, pp. 901 – 904. ISSN : 0975-3397 1835

[26] Y.Wang, G. Attebury, and B. Ramamurthy, "A Survey of Security Issues in Wireless Sensor Networks" IEEE Communication. Survey Tutorials, vol. 8, pp. 2–23.

[27] Y.Mun and C. Shin, "Secure routing in sensor networks: Security problem analysis and countermeasures," in International Conference on Computational Science and Its Applications - ICCSA 2005, vol. 3480 of Lecture Notes in Computer Science, (Singapore), pp. 459–467, Springer Verlag, Heidelberg, D-69121, Germany.

[28] Thomas Haenselmann. Sensornetworks. GFDL Wireless Sensor Network textbook

[29] Secure and Efficient Broadcast Authentication in Wireless Sensor Networks Taekyoung Kwon, Member, IEEE, and Jin Hong.

[30] World Academy of Science, Engineering and Technology 51 2009 Secure Data Aggregation Using Clusters in Sensor Networks Prakash G L, Thejaswini M, S H Manjula, K R Venugopal, L M Patnaik

9

A PROTOCOL TO IMPROVE THE DATA COMMUNICATION OVER WIRELESS NETWORK

S Saravanan[1] and E Karthikeyan[2]

[1]Research Scholar, Dept of Computer Science, Bharathiar University, Coimbatore, India
`ssam2020@gmail.com`
[2]Asst. Professor of Computer Science, Government Arts College, Udumalpet, India
`e_karthi@yahoo.com`

ABSTRACT

Reliable transport protocols are tuned to perform well in traditional networks where packet losses occur mostly because of congestion. However, in networks with wireless links in addition to wired segments this assumption would be insufficient, as the high wireless bit error rate could become the dominant cause of packet loss. The main reason of this poor performance for TCP is that TCP cannot distinguish between packets losses due to wireless errors from those due to congestion. Moreover, TCP sender cannot keep the size of its congestion window at optimum level and always has to retransmit packets after waiting for timeout, which significantly degrades throughput and end-to-end performance of TCP.

In this work, a novel protocol, called DLN (Data Loss Notification), is be proposed. By changing the ACK format of TCP, we could successfully distinguish packet losses incurred by congestion and channel error which will improve both throughput and delay performance of TCP in wireless environment significantly. This mechanism has been demonstrated to provide performance improvements across a range of bit-error rates.

KEYWORDS

Data Loss Notification, Wireless Error, Network Congestion, Loss Recovery, Throughput

1. INTRODUCTION

TCP is a protocol developed on wired Internet, some of the algorithm is based on characteristics of wired network. One most important aspect is that TCP host may think packet loss as network congestion. This assumption is pretty reasonable in wired Internet, because the wired transmission media has quite a small packet loss ratio; the main reason for packet loss is network congestion.

The thing is not that right in the rapid developing wireless network. The media of wireless network has totally different characteristics. The wireless channel is an error prone media; the packet transmitted through wireless link may be lost due to wireless error. The TCP host may judge the packet lost due to network congestion, and then take the action of reducing size of the transmission window or waiting for long idle time for retransmitting the lost packet. The result is degraded end-to-end performance. In addition, packet losses that occur due to use mobility cause the TCP sender to remain idle for long periods of time even after the handoff is completed, resulting in unacceptably low throughput.

Link layer protocols are an alternative for improving the poor performance of TCP over wireless link. In those methods usually forward error connection (FEC) or automatic repeat request (ARQ) are used to improve the performance. Independent timer reaction at link and transport layers that may result in unnecessary retransmission, fast retransmission interaction, and large round-trip variation are considered as major problem with link-layer approaches. Split-connection protocols attempt shield the sender from the wireless link by explicitly terminating the wired connection at the base station and using a separate transport connection over the wireless link. However, they do not preserve end-to-end semantics because data may be acknowledged to the sender even before it reaches the receiver, complicate handoff procedures because they involve hard state in the network, and do not usually provide the best possible performance. Another enhancement to TCP for wireless channel reviewed here is called Snoop protocol. In this method, the base station is equipped with a module called snoop agent, which its function is to monitor the TCP packets transmitted from a fixed host to a mobile host and vice versa. The agent caches all those packets locally and in the case of receiving duplicate acknowledgements (ACKs), retransmits the packets promptly and suppresses duplicate ACKs. The Snoop protocol performs retransmission of lost packets locally (at the base station) and hence avoids lengthy fast retransmission and congestion control at the sender side. By this method, end-to-end semantics of TCP is maintained and performance of TCP is improved. The Snoop protocol is mainly used for the fixed host to mobile host direction, explicit loss notification algorithms complementing the Snoop on the mobile host to fixed host direction.

Our approach, called the Data Loss Notification (DLN) protocol, explicit optimization to improve performance. We design a mechanism that can successfully distinguish between congestion and channel error-induced packet losses to substantially enhance end-to-end performance. The DLN protocol uses only soft state at agent in the network, which is periodically refreshed upon the arrival of data segments and ACKs.

Another aspect of Internet packet we analyse in this paper is that of packet delay. Delay variation is arguably the most complex element of network behaviour to analyse with loss, for example, the packet either shows up at the receiver or it does not, while with delay there are many shades of possibility and meaning in the time required for a packet to arrive. Likewise, delay variation is potentially the richest source of information about the network.

Small TCP transmission window prevent the sender from recovering from losses without incurring expensive timeouts that keep the connection idle for long periods of time. As a result, Delay for packet from sender host to receiver host is high due to the time for waiting for timeout and retransmission. Thus, a key challenge is in enhancing TCP's loss recovery algorithms when large wireless packet loss rate leads to numerous timeouts.

Based on an extensive analysis, we show that not only are current loss recovery techniques grossly inadequate at preventing sender timeouts, but that proposed enhancements like Selective Acknowledgements (SACK) are not likely to significantly change this because typical Internet transmission window are not larger than a few segments. Based on the results of our analysis, the DLN has the function of judge the reason of packet loss, which can efficiently retransmit lost packet without unnecessarily reducing the transmission window.

2. BACKGROUND AND RELATED WORK

The purpose of this section is to give an introduction of reliable transmission protocol and the application of reliable data transmission in wireless network. We begin in section 2.1 from discuss how reliable transport protocol developed, and then give a detailed description of various transport protocol discussed in the literature. In section 2.2, we give an introduction of Transmission Control Protocol, which is the main protocol used in the Internet. In section 2.3, we discuss several end-to-end TCP protocols developed in these years.

2.1 Reliable transport Protocols

Today's Internet is a best effort transmission network. Packet is transmitted in the network without guarantee of in order and reliable transmission, this is a big difference compared with traditional telephone transmission. The transmission host sends out the packet into the network, then the packet passes through a series of routers to its destination, these routers determined the routing of the packet. Packet may be lost due to network congestion. While this architecture of Internet ensured a simple and effective Internet protocol, which expedited the rapid grow of Internet. It sends the mission of reliable packet transmission to the higher layer of protocols. Applications like the World Wide Web [1], file transfer [2], remote terminals and electronic mail [3] that need reliable and ordered data delivery require a transport protocol to provide this functionality, freeing them of the need to achieve reliability on a per-application basis.

Several reliable transport protocols have been proposed in the literature for best-effort networks: Delta-t [4], NETBLT [5], VMTP [6], OSI/TP4 [7], XTP [8] and TCP [9]. In all these protocols, data items are identified by sequence numbers that are either byte-based or packet-based. These sequence numbers are used to detect losses, reordering and data duplication. Packet-based sequence numbers are simple to implement but are less flexible because they often constrain the sender to use fixed-size packets. Byte-based sequence numbers indicate to the receiver the exact amount of missing data and permit the sender to precisely identify and retransmit lost data.

2.2 Transmission Control Protocol (TCP)

In the Internet today, TCP is now the standard for reliable data transport. Measurements made in 1999 show that over 95% of all bytes, 90% of all packets and 75% of all flows use TCP. This section discusses its salient features and highlights its main weaknesses.

While the original formal specification of TCP is in RFC793, numerous variants of it have been developed over the past several years, such as TCP-Tahoe, TCP-Reno, Vegas etc. This section discusses the TCP-Reno variant of TCP, which is the predominantly deployed version today.

2.2.1 Cumulative Acknowledgments

TCP is an ARQ-based reliable transport protocol that uses cumulative ACKs and byte based sequence numbers for reliability. TCP provides a fully reliable, in-order; byte-stream delivery abstraction to the higher-layer application, which typical uses a socket interface to interface with the transport layer. The basic unit of transmission is called segment, which is a contiguous sequence of bytes identified by its 32-bit long start and end sequence numbers. The transmitted segments are smaller than or equal to the connection's maximum segment size (MSS), which is negotiated at the start of the connection.

2.2.2 Loss Recovery

When the TCP sender discovers that data has been lost in the network, it recovers from it by retransmitting the missing segments. TCP has two mechanisms for discovering and recovering from losses: timer-driven retransmissions and data-driven retransmissions.

Time-driven recovery: When the TCP sender does not receive a positive cumulative ACK for a segment within a certain time-out interval, it retransmits the missing data. To determine the timeout interval, it maintains a running estimate of the connection's round-trip time using an exponential weighted moving average (EWMA) formula, srtt=a*rtt+(1-a)*srtt, where 'srtt' is the smoothed round-trip time average, rtt is the current round-trip sample, and 'a' the EWMA constant set to 0.125 in the TCP specification. It also estimates the mean linear deviation, 'rttbvar', using a similar EWMA filter, with a set to 0.25. A time-out occurs if the sender does not receive an ACK for a segment within 'srtt+4*rttvar' since the arrival of the last new cumulative ACK. Furthermore, the retransmission timer is exponentially backed off after each unsuccessful retransmission. The details of the round-trip time calculations and timer management can be found in [10,11,12].

Data-driven recovery: TCP's data-driven retransmission mechanism uses a technique called Fast Retransmission. It relies on the information conveyed by cumulative ACKs and takes advantage of the receipt of later data segments after a lost one. Because ACKs are cumulative, all segments after a missing one generates duplicate cumulative ACKs that are sent to the TCP sender. The sender uses these duplicate ACKs to deduce that a segment is missing and retransmits it.

However, the sender must not retransmit a segment upon the arrival of the very first duplicate ACK. This is because the Internet service model does not preclude the reordering of packets in the network; such reordering cause's later segments to be received ahead of earlier ones, and triggers duplicate ACKs in the same way that losses do. Furthermore, the degree of packet reordering on the Internet seems to be increasing, thus, to avoid prematurely retransmitting segments, the sender waits for three duplicate ACKs, the current standard fast retransmit threshold.

This is followed by the fast recovery phase, where additional packets are transmitted after the sender is sure that at least half the current window has reached the receiver, based on a count of the number of received duplicate ACK. Fast recovery ensures that a fast retransmission is followed by congestion avoidance and not by slow start. Since the arrival of duplicate ACKs signals to the sender that data is indeed flowing between the two ends, there is no reason to suddenly throttle the sender by invoking slow start.

2.2.3 Congestion Avoidance and Control

TCP's congestion management is based largely on Jacobson's seminar paper [10]. TCP uses a window-based algorithm to manage congestion, where the window is an estimate of the number of bytes currently unacknowledged and outstanding in the network.

The TCP sender performs flow control by ensuring that the transmission window does not exceed the receiver's advertised window size. It performs congestion control by using a window-based scheme, where the sender regulates the amount of transmitted data using a congestion window. When a connection starts or resumes after an idle period of time, slow start is performed. Here, the congestion window is initialized to one segment and every new ACK increases the window by one MSS. After a certain threshold (called the slow start threshold, 'ssthresh') is reached, the connection moves into the congestion avoidance phase, in which the congestion window effectively increases by one segment for each successfully transmitted window. In response to a packet loss, the sender halves its congestion window; if a timeout occurs, the congestion window is set to one segment and the connection goes through slow start once again.

2.3 End–to-End TCP Enhancements

Over the past decade, TCP has been tuned to work well in wired networks in the face of network congestion, reacting to congestion by reducing its rate, recovering from lost segments, and probing for bandwidth in a careful way. In this section, we survey some proposed enhancements to TCP that improve its ability to recover from losses in a timely manner and/or perform better congestion control.

2.3.1 Selective Acknowledgments (SACK)

It is well known that TCP performance suffers due to coarse-grained timeouts when multiple segments are lost in a single window because it uses only cumulative ACKs. There has therefore been recent interest in adding selective acknowledgements (SACK) to the standard TCP specification to reduce the time it takes to recover from multiple losses in a window.

TCP Selective Acknowledgements can be implemented in many ways. One possible approach is to use Keshav and Morgan's SMART (Selective Mechanism to Aid Retransmission) scheme, where the receiver communicates the segment number that just arrived in addition to the cumulative ACK, whenever it sends a duplicate ACK. A second approach, described in RFC2018, is currently on track to become an Internet standard. In this scheme, the receiver reports up to three of the last received, out-of-order, maximal contiguous blocks of data, in addition to the cumulative ACK, so that the sender can accurately deduce which segments have reached the receiver.

2.3.2 NewReno

Hoes' NewReno modification to TCP-Reno reduces the number of timeouts incurred by the TCP sender when multiple losses happen in a transmission window [13]. When multiple losses occur in TCP-Reno, a fast retransmission occurs after three duplicate ACKs arrive. Later duplicate ACKs are ignored until half the window is acknowledged, after which fast recovery sends a new segment for every incoming duplicate ACK. Now, when a new ACK arrives after the successful fast retransmission, its value will be within the original window (recall that there were multiple losses in the original window). In many cases TCP-Reno would time out for this second loss if the second loss is "close" to the location of the first one, because the sender's window would now have been reduced to half its original value and the sliding window not have shifted the original window to beyond the original right edge.

In NewReno, however, the sender remains in fast recovery when this happens, when the new ACK arriving after a fast retransmission is partial. A partial ACK is defined as a cumulative ACK that does not acknowledge the original window completely. By remaining in fast recovery, the sender continues to send new segments, which would elicit more duplicate ACKs and eventually trigger another fast retransmission without incurring a timeout.

2.3.3 Forward Acknowledgments

Mathis and Mahdavi's TCP with forward Acknowledgments (FACK) [14] uses SACK information at the sender to perform better congestion management. Rather than assume that the receiver on every duplicate ACK has received a MSS-worth of data, FACK calculates this precisely using information from the SACK field. It explicitly maintains an "available window" variable, awnd, that keeps track of the number of bytes that are unacknowledged, to perform better congestion control. As long as awnd is smaller than cwnd, the sender is permitted to transmit more data.

2.3.4 NetReno

Concurrent with our work, Lin and Kung [15] discuss some mechanisms for improving TCP loss recovery and propose some modifications to TCP. They argue that their modifications are more sensitive to network conditions than current TCP is motivating the name NetReno (for "Network Sensitive Reno").

2.3.5 Snoop Protocol

Snoop protocol improves TCP performance by deploying an agent at the base station. The agent mainly performs the functions of loss detection and loss recovery via retransmission by taking advantage of the information conveyed in TCP acknowledgments (ACKs) from the receiver. For transfer of data from a fixed host to a mobile host, the snoop agent used the loss indications conveyed by duplicate TCP ACKs and locally maintained timers to retransmit loss data from the base station. The agent also suppresses duplicate ACKs corresponding to wireless losses from the TCP sender, Thereby preventing unnecessary congestion control invocations.

Snoop protocol can suppress duplicate acknowledgment for TCP segments lost and retransmitted locally, thereby avoiding unnecessary fast retransmission and congestion control invocations by the sender. Also there are some disadvantages with it: the agent must be TCP-aware, as a result, this scheme is protocol dependent and cannot work for other existing protocols or future protocols when they become available. Although a lost packet can be retransmitted locally by the base station, the generated three duplicated acknowledgment packets still reach the sender of the TCP connection and cause the sender to unnecessarily reduce it sending rate by 50%.

2.3.6 Explicit Congestion Notification

Explicit congestion Notification is provided by Internet router for indication of incipient congestion where the notification can sometimes be through marking packets rather than dropping them. This would require an ECN field in the IP header with two bits. The ECN-capable Transport (ECT) bit would be set by the data sender to indicate that the end-points of the transport protocol are ECN-capable. The bit would be set by the router to indicate congestion in the end nodes. Routers that have a packet arriving at a full queue would drop the packet, just as they do now.

When gateway en route is congested or close due to congestion, it sets a bit in the packet header and Forward it on. A Random Early Detection (RED) gateway in marking mode is apt for this. A RED gateway detects incipient congestion by tracking the average queue size over a time window in the recent past. If the average exceeds a threshold, the gateway selects a packet at random and marks it, by setting the Explicit Congestion Notification (ECN) bit in the IP packet Header. This notification is echoed to the sender of the packet by the receiver. For TCP, This echo is piggybacked on an ACK. Now when the sender receives ACK with ECN, it reduces its congestion window, as it would if a packet loss had occurred. Thus this mechanism with explicit support from gateways allows TCP to perform proactive congestion control, over and above the reactive one triggered by packet loss [16].

The ECN algorithm can avoid depending on packet drops alone as implement congestion avoidance mechanisms.

3. PROPOSED DLN METHOD

3.1 Protocol Description

The algorithms proposed in this paper tried to improve the performance of TCP in wireless networks. But none of these algorithms actually lets TCP sender know clearly whether the packet is lost due to wireless error or network congestion. This makes the TCP sender retransmits the packet efficiently and then cannot keep the throughput high in the error prone environment.

The Snoop protocol is a good scheme to improve the performance of TCP in wireless network. But the Snoop protocol retransmits the lost packet like other link layer solutions. The Snoop protocol also suffers from not being able to completely shield the sender from the wireless losses. Based on Snoop protocol, we proposed a new protocol called Data Loss Notification (DLN) which can remedy limitations of the Snoop protocol.

In order to implement DLN protocol we modify the ACK format and the networking software at the base station and the mobile host.

3.1.1 DLN structure

We use a new form of ACK named DLN. In DLN we add the sequence number of the four most recently lost packets judged by the mobile host for each lost packet, and one bit (called

DLN bit) to indicate the reason of the lost packet. 1 indicates the packet is lost in the wired network congestion and 0 indicates the packet is lost due to wireless error. The default value of DLN bit transmitted by mobile host is 1 (assuming the corresponding packet was lost due to network congestion). The DLN bit is judged at the base station. DLN agent at the base station checked the information it stored in it to see if the packet has lost before it is arrived in the base station. If it found the packet had lost before it arrived in the base station, it remained the corresponding DLN bit to 1, or it would remain the DLN bit to 0. After the DLN is processed by DLN agent at the base station, it continues to be transmitted back to the fixed host. When the fixed host received the DLN, it would know the reason of packet loss from the DLN bit.

3.1.2 DLN agent at the base-station

The DLN agent at the base station has two main functions. One is to judge and store the packet lost information transmitted from fixed host. Like ordinary wired network, packet transmitted from fixed host to base station may be lost due to congestion. If the base station does not receive the packet in sequence, it will store the corresponding packet information in the DLN agent. Using the information stored in the agent, the base station can judge the reason of packet loss when it receive the DLN transmitted from mobile host. The second function is to judge the value of DLN. When the base station receives DLN, it will judge the lost packet based on the information it stored. If it finds the lost packet is lost before it arrived in base station, it will fill the DLN bit with 1 to indicate the packet was lost due to congestion. If the lost packet had arrived in the base station, it fills the DLN bit with 0 to indicate the packet was lost in the wireless channel.

3.1.3 Fixed host TCP sender

When the fixed host receives the DLN, it actions with the information contained in the DLN bit. If the DLN bit is 1, means the corresponding packet is lost due to wired congestion; TCP sender will proceed same as the window algorithm. If the DLN bit is 0, it means the corresponding packet is lost due to wireless error and it retransmits the packet immediately without window reduction.

3.2 Model

The DLN agent maintains a cache of TCP packets sent from the fixed host that have been forwarded to, but not yet been acknowledged by the mobile host. This is easy to do since TCP has a cumulative ACK policy. When a new packet arrives from the fixed host, the agent adds it to its cache and passes the packet on to the forwarding code, which performs the normal packet forwarding functions. The DLN agent also monitors TCP ACKs sent from the mobile host, using new ACKs to clean its cache and maintain only unacknowledged packet there.

The DLN agent has two main components: data processing and ACK processing. The flowcharts summarizing the salient features of the algorithms shown in Figure 1 and Figure 2. The remainder of this section describes the algorithm in detail.

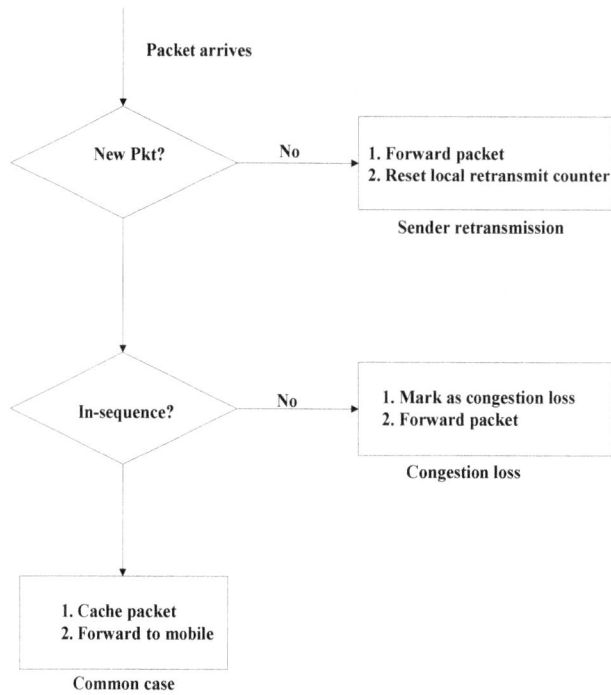

Figure 1. Flowchart for data processing in DLN

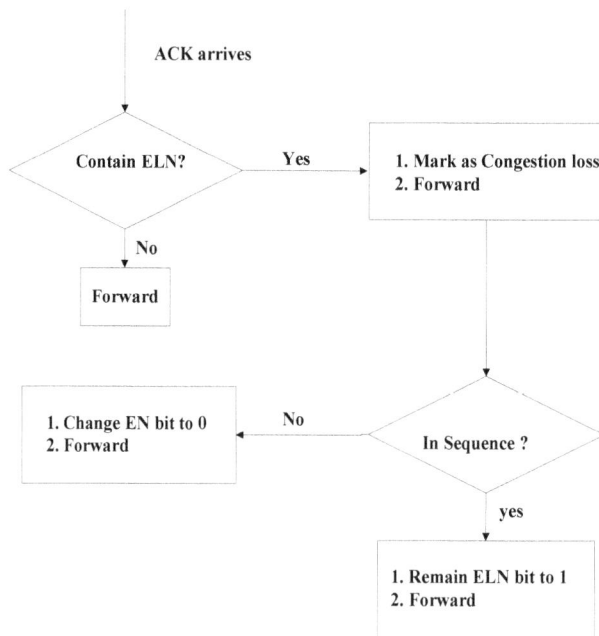

Figure 2. Flowchart for ACK processing in DLN

3.2.1 Data Processing

In this phase, the DLN agent processes incoming data segments from the fixed host. A TCP segment is identified by the sequence number of its first byte of data and its size. At the base station, the DLN agent keeps track of the last sequence number seen for the connection. At any stage in the protocol, one of the several kinds of packets can arrive at the base station from the fixed host, and the DLN agent processes them in different ways:

A new packet in the increasing TCP sequence: This is the common case, when a new packet in the normal increasing sequence arrives at the base station. In this case the packet is added to the DLN agent cache and forwarded on to the mobile host. The agent does not perform any extra copying of data while doing this.

An out-of-sequence packet that has been cached earlier: Although this is an uncommon case, it can happen. If there have been multiple losses in a single window due to congestion on the wired path, there are times when a timeout and slow start occur after a fast retransmission. This could lead to a sender retransmission of a previously cached segment. If the sequence number is greater than the last ACK seen, it is very likely that this packet didn't reach the mobile host earlier, and so it is forwarded on. If, on the other hand, the sequence number is less than the last ACK, the mobile host has probably already received this packet. At this point, there are several possible actions the agent could take. One possibility would be to discard this packet and continue, but this is not always the best thing to do. The reason for this is that the original ACK with the same sequence number could have been lost due to congestion while going back to the fixed host. The second possibility, to facilitate the sender getting to the current state of the last ACK seen at the base station (with the source address and port corresponding to the mobile host) to the fixed host. However, the disadvantage of this is that if the information in the DLN agent's state is wrong for any reason, the correctness of the end-to-end protocol is compromised. The third option is to simply forward the packet to the mobile host, and await information from subsequent ACKs to refresh the state at the agent. This is the option the agent uses, in keeping with our soft-state philosophies. The agent also resets the number of local retransmissions to zero, and updates the transmission time of this segment to correctly estimate the round-trip time when its ACK arrives.

An out-of-sequence packet that has not been cached earlier: In this case the packet was either lost earlier due to congestion on the wired network, or has been delivered out-of-order by the network. The former is more likely, especially if the sequence number of the packet (i.e., the sequence number of its first data type) is more than one or two packets away from the last one seen so far by the DLN agent. The agent and then forward to the mobile host cache this packet. It is also marked as having been retransmitted by the sender. The DLN processing algorithm uses this information to process DLN bit contained in ACKs that from the mobile host.

3.2.2 DLN Processing

The mobile host produces DLN if there is lost packet information contained in the receiving packet sequence. Each DLN packet may contain at most three lost packets that are most near by the acknowledged packet. Because the mobile host cannot judge if the packet is lost due to congestion or wireless channel error, it always fills the DLN bit corresponding to each lost packet for 1 (assume the packet is lost due to wired network congestion). In the base station when DLN agent receive DLN packet, it can use the information stored in the cache to judge the reason of the packet loss.

The DLN agent performs various operations depending on type of ACKs it receives. These ACKs fall into one of the following categories:

A new, expected ACK: This is the common case that occurs when no recent segments have been lost, and signifies an increase in the packet sequence received at the mobile host. This

ACK initiates the cleaning of the DLN agent cache and all acknowledged segments are freed. Finally, the ACK is forwarded to the fixed host.

An ACK contain DLN information: This is an ACK that contains one or several lost packet information. The lost packet is the packet that was not received by the mobile host. When the DLN agent received the ACK, it checks the information of data transmitted from fixed host. If the lost packet contained in the DLN is also cached in the agent that means this packet was lost before it was transmitted to the base station, the DLN agent marks the DLN bit of the corresponding packet to 1. If the lost packet contained in the DLN is not cached in the agent, that means this packet was lost after the base station, then it marked the DLN bit of the corresponding packet to 0.

A spurious ACK: This is an ACK less than the last ACK seen by the DLN agent, and is a situation that rarely occurs. We forward it to the fixed host, to guard against the possibility that the internal state of the DLN agent may be incorrect.

3.3 Simulation and Performance Results

We performed several simulations with the DLN protocol and compared the resulting performance with other algorithms using well known simulator ns-2. We present the results of the experiments in this section. We first introduce the simulation topology and parameters used in the simulation in Section 3.3.1. Later we discuss a trace comparison of Reno with DLN in high wireless error environment in Section 3.3.2 and the results of detailed simulation on different TCP algorithm in different parameters in Section 3.3.3. In section 3.3.4 we give simulation results of delay and window size in the Reno and DLN procession, and an analysis of the simulation results.

3.3.1 Simulation Topology

Figure 4 shows the network used for the simulations in this paper. The circle indicates a finite-buffer drop-tail gateway, and the squares indicate sending and receiving hosts. In the simulation, some parameters can be set to indicate different network condition; the parameters are summarized as following: 1. Buffer size (B packets) in the base station, 2. Propagation delay (D msec) which includes: 1) the time between the release of a packet from the source and its arrival into the link buffer; 2) the time between the transmission of the packet on the bottleneck link and its arrival as its destination; and 3) the time between the arrival of the packet at the destination and the arrival of the corresponding acknowledgment at the source, 3. The bandwidth (U packets/msec) of bottleneck link from base station to mobile host.

Figure 4. Simulation Topology

3.3.2 Trace Comparison

Figure 5 shows the sequence traces of TCP transfers over an error-free link, as well as TCP-Reno and the DLN protocol over error-prone links. The middle curve shows the progress of a transfer using the DLN protocol, which is a curve close to the ideal, error-free case. The performance improvement in this transfer over TCP-Reno is a factor of four. This is typical of the degree of improvement we obtain in our ns-2 simulation for reliable data transfers.

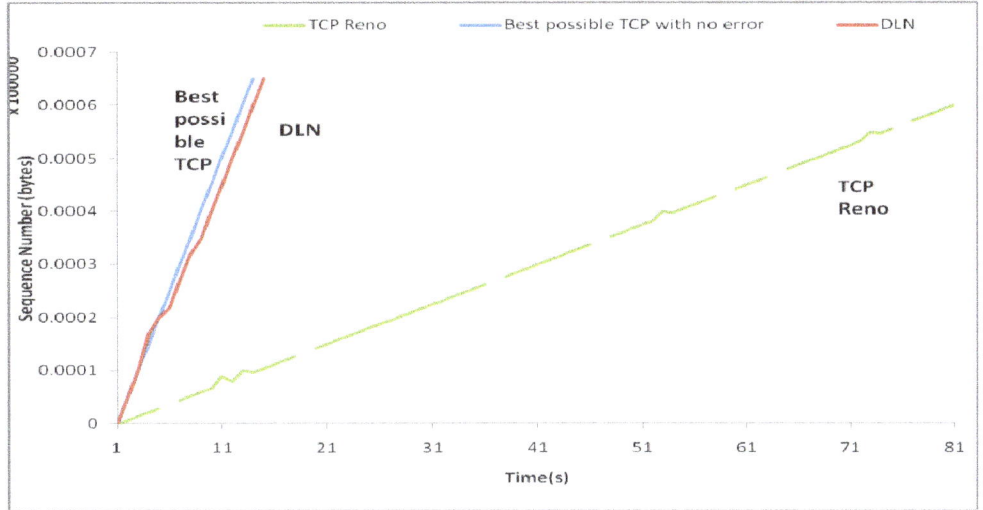

Figure 5. Sequence traces of TCP using an Ideal error-free links, as well as TCP Reno and the DLN protocol over error prone links

3.3.3 Throughput Comparison

Figure 6 to 9 show the throughput of algorithms under different wireless packet loss rate. We perform simulation with different network condition by changing the parameters such as buffer size, propagation delay and the bandwidth of bottleneck link. From simulation results we can see that throughput of TCP-Reno and TCP-Tahoe drop sharply when the error rate is increased to above 10-2. Throughput of TCP-Reno and TCP-Tahoe drop to only 10%~20% compared with no error rate in wireless link, DLN scheme can keep the throughput as high as 80%~90% of error free environment. There are significant performance benefits of using the DLN protocol. The main advantage of DLN is that it helps maintaining a large TCP congestion window when wireless error rate is high.

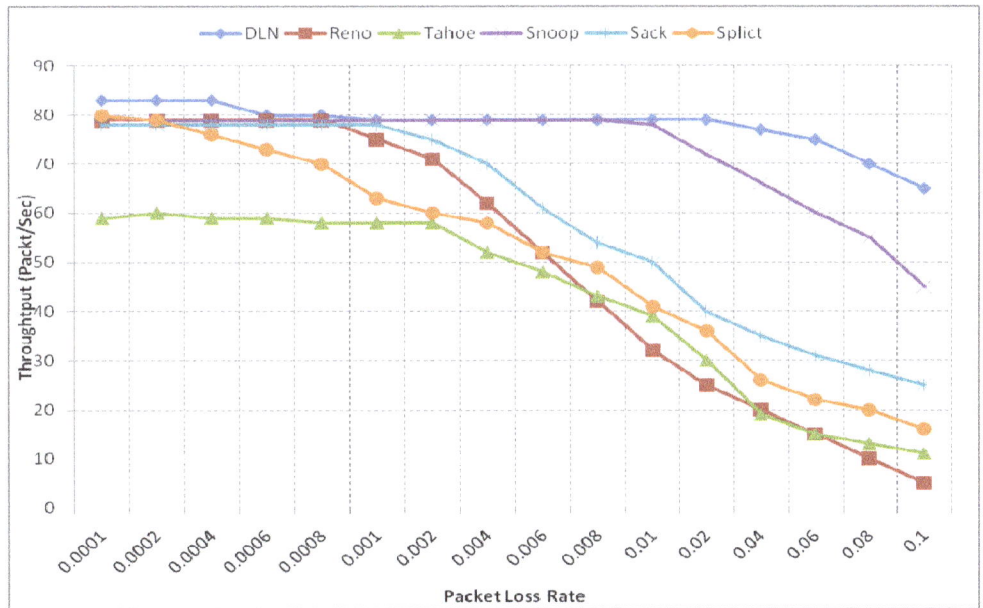

Figure 6. Simulation comparison (B=5; D=0.2; U=100)

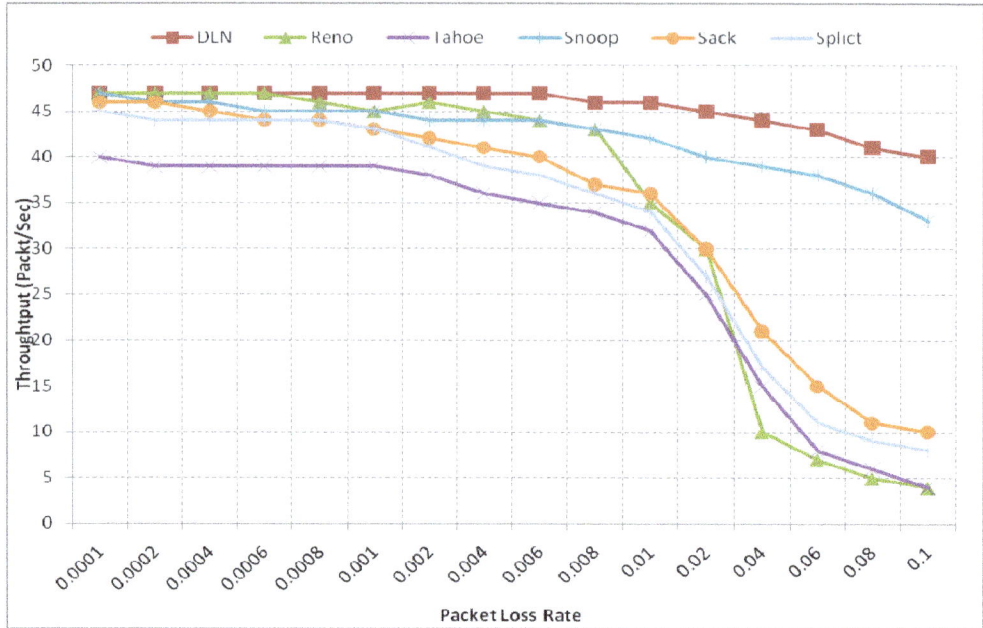

Figure 7 Simulation comparison (B=5; D=0.2; U=50)

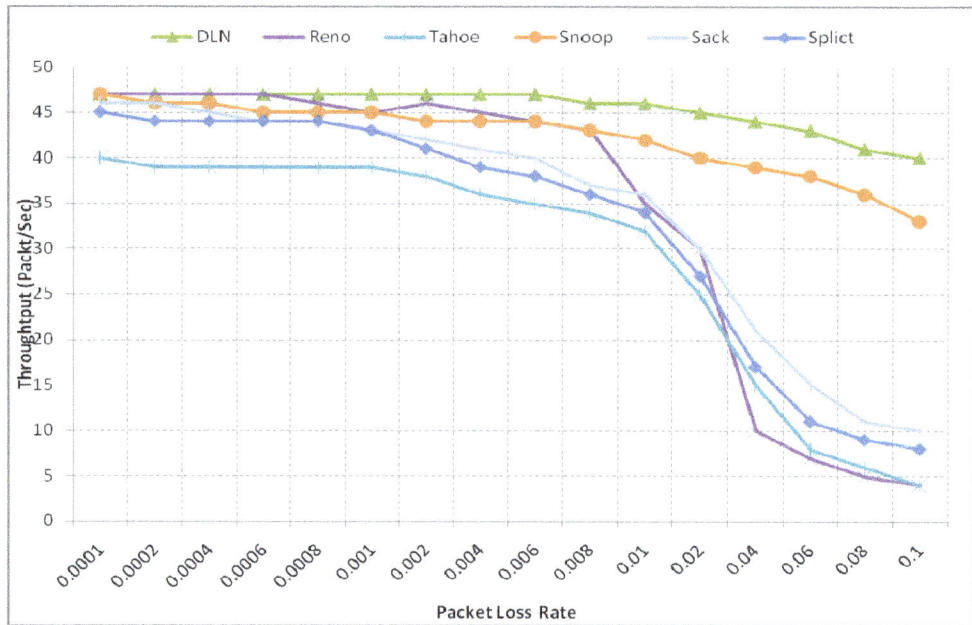

Figure 8 Simulation comparison (B=8; D=0.2; U=50)

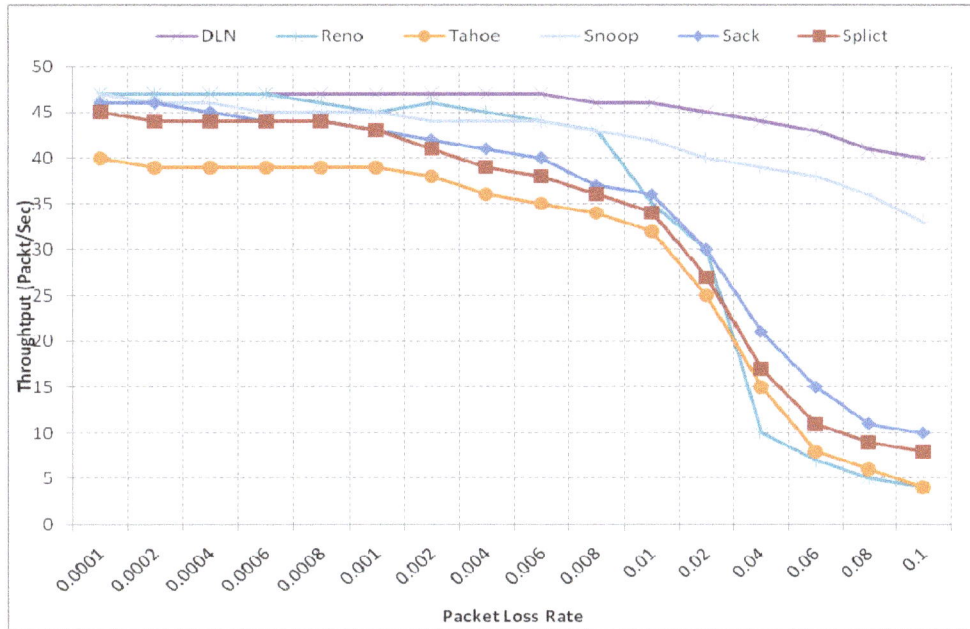

Figure 9 Simulation comparison (B=5; D=0.1; U=50)

3.3.4 Delay and Loss Recovery Analysis

In this section, we analyze the problems that arise due to small TCP transmission windows and long delay and discuss the function of DLN on solving this kind of problem [17].

TCP-Reno is the most popular TCP used in the Internet. It contains a number of algorithms aimed at controlling network congestion while maintaining good user packet is dropped for a window of data, but can suffer from performance problems when multiple packets are dropped due to the wireless error. In this section we illustrate the problem of TCP-Reno, show the end-to-end performance degradation of TCP-Reno in high packet loss wireless channel. We also compare the simulation results of the TCP-Reno algorithm with DLN to show the performance improvement of DLN.

If the packet is successfully transmitted from sender host to the receiver, the end-to-end delay is mainly determined by propagation delay, service time and queuing at the base station. But if the packet is lost due to network congestion or wireless packet loss, the TCP sender has to retransmit the lost packet by performing a loss recovery task and by waiting for time out. As a result, the end-to-end delay becomes significantly long when timeout happens.

Figure 10 to 12 show the delay of 200 packets transmission using TCP-Reno and DLN.

The TCP-Reno has quite good performance when there is no wireless error in the wireless channel. Figure 12 shows the end-to-end delay performance of TCP-Reno in error free environment. The mean end-to-end delay is about 0.15~0.2 second. There are two packets with delay around 0.5 second, this is due to network congestion, and these two packets are retransmitted by loss recovery mechanism.

In Figures 10 and 11 the packet loss rate in wireless link is 0.1, which means in 200 packets transmission there are about 20 packets lost in the wireless channel. Based on the simulation result, in TCP-Reno, the mean transmission delay is about 0.15~0.25 second and we can see that in 200 packets transmission, there are 24 packets whose transmission delay is significantly above 0.2 second. These packets are lost somewhere (because of network congestion or wireless error) in the network. Of the 24 retransmitted packets, 11 packets have the delay around 0.4 to 0.6 second; that means these packets are retransmitted by loss recovery mechanism without invoking timeout, 13 other packets have a delay around 1 to 1.4 seconds, which means these packets are time out. The TCP sender always has to wait for time out to retransmit the lost packet, due to wireless packet loss. The DLN algorithm can efficiently avoid the timeout by retransmitting the lost packet immediately. Figure 12 we can see that the mean end-to-end delay for packet transmission is about 0.1~0.2 seconds. There are 23 packets with end-to-end delay for packet transmission is about 0.1~0.2 seconds. There are 23 packets with end-to-end delay between 0.4 to 0.6 seconds. Most of these packets are lost in wireless channel when it is first transmitted by the TCP-sender. Unlike TCP-Reno, the TCP sender knows these packets are lost due to wireless error but not network congestion. The lost packet can be retransmitted efficiently without incurring any window deduction, which avoided long idle time to wait for timeout.

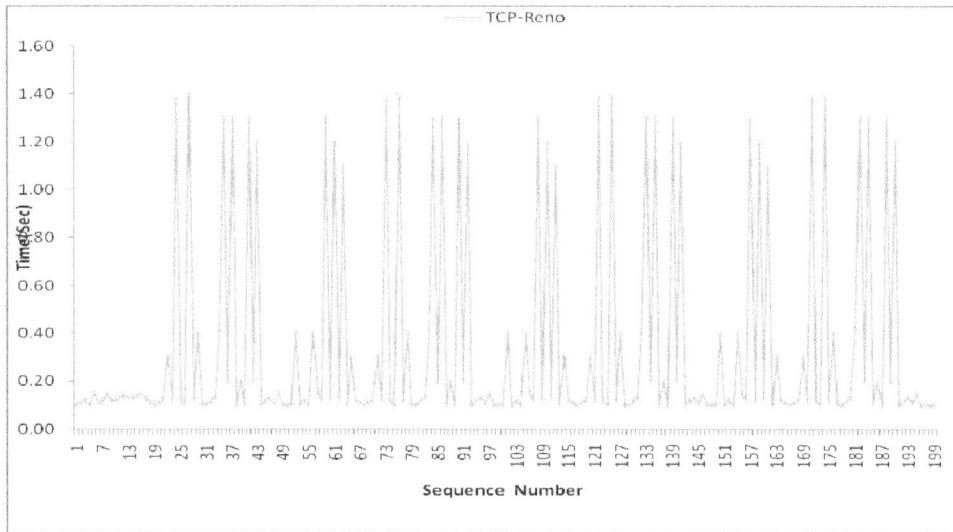

Fig 10. End-to-end delay for TCP-Reno (Wireless packet loss rate = 0.1)

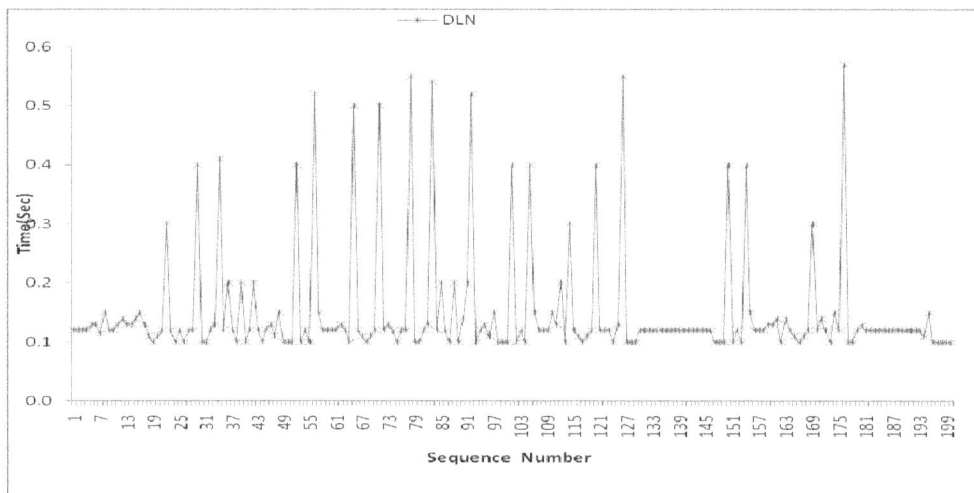

Fig 11. End-to-end delay for DLN (Wireless packet loss rate = 0.1)

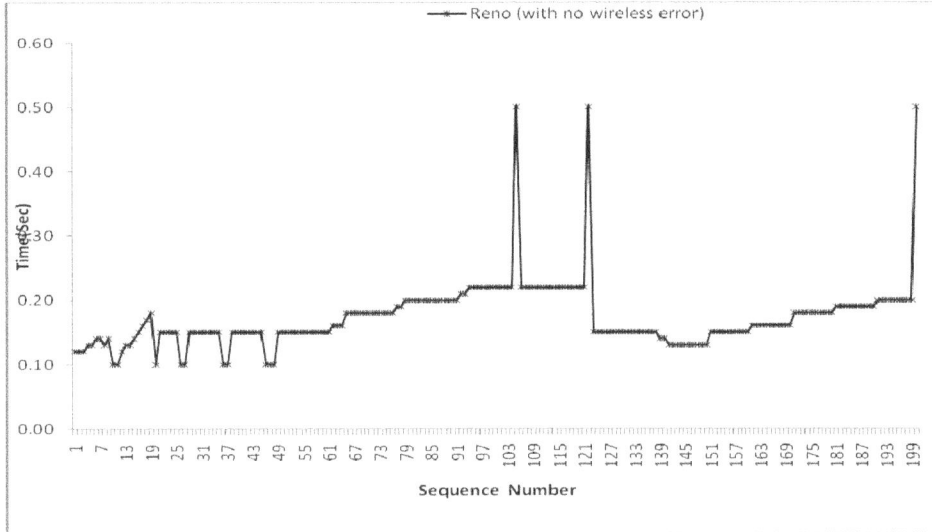

Fig 12. End-to-end delay for TCP-Reno (no wireless error)

The main reason for the occurrence of heavy timeouts in the TCP-Reno algorithm is the small congestion window, which makes TCP sender not to get enough duplicated acknowledgement in the procession and that the number of arriving duplicate ACKs is not sufficient to trigger a fast retransmission. The result is a timeout-driven transmission that keeps the link idle for long periods of time. Figure 13 to 15 show the congestion window size in the procession of transmitting 200 packets. From Figure 13 we can see the window cannot open big enough and always reduce to one due to timeout. So in the high packet loss environment, the TCP-Reno cannot efficiently transmit packet.

When there is no wireless packet loss, the only packet loss in transmission is due to network congestion. The TCP-Reno sender retransmits packet by using loss recovery algorithm. The window size is reduced to half when loss recovery happens, but no time out happens, because the congestion window (shown in Figure 15) is kept big enough and there are enough duplicated ACKs transmitted back to trigger the loss recovery.

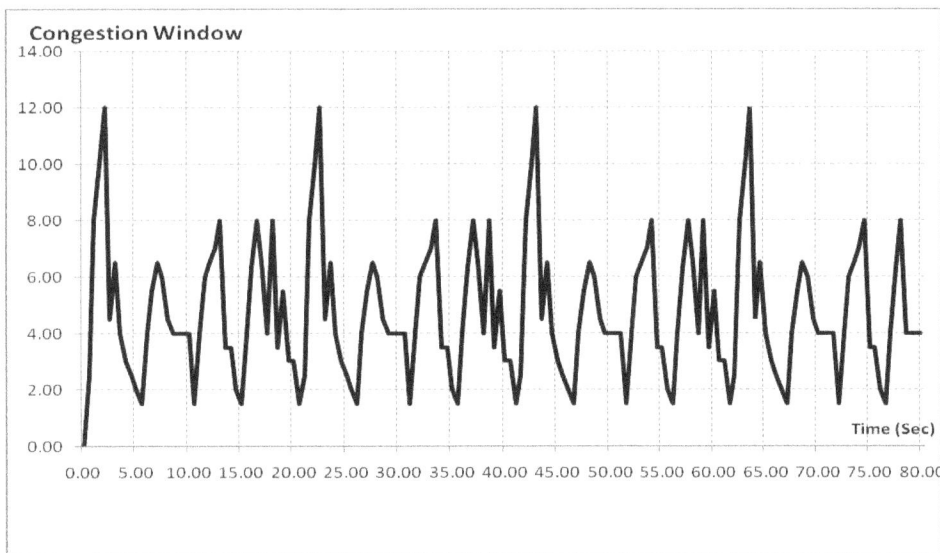

Fig 13. Window evolution for TCP-Reno (wireless packet loss rate=0.1)

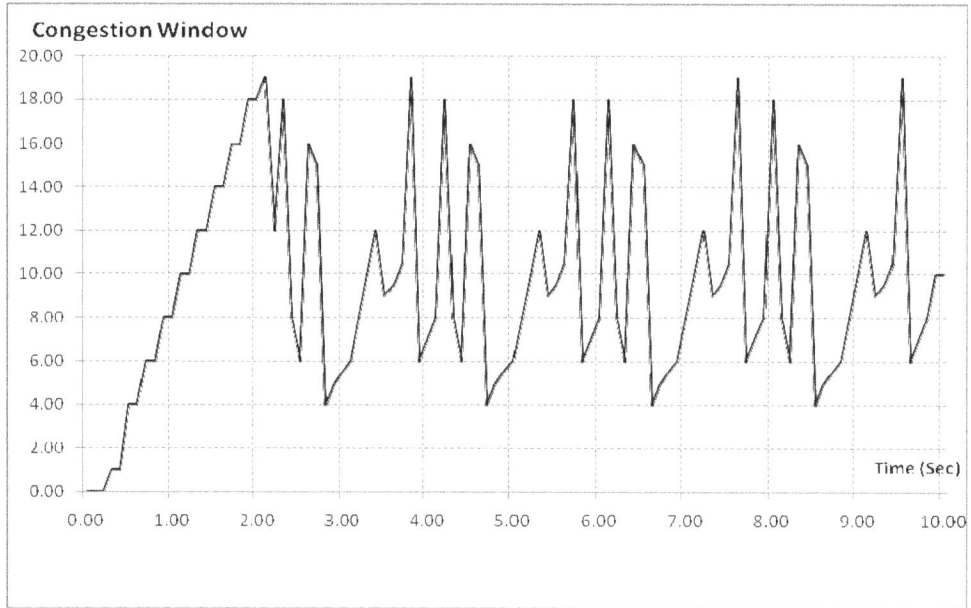

Fig 14. Window evolution for DLN (wireless packet loss rate=0.1)

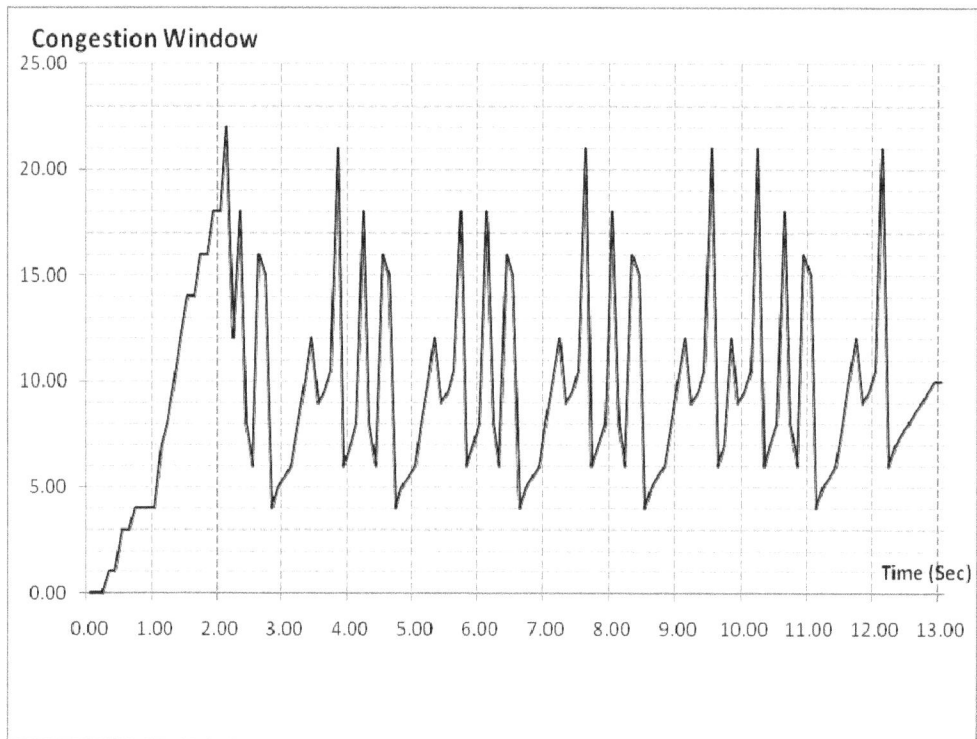

Fig 15. Window evolution for TCP-Reno (no wireless error)

4. CONCLUSION

In this paper, we identified fundamental challenges to improve the performance of TCP in wireless networks. TCP performance in many wireless networks suffers because bit-error-induced packet losses, which occur in burst because of the nature of the wireless channel, are misinterpreted by the TCP sender. TCP attributes these losses to network congestion because of the implicit assumptions made by its congestion control algorithms today. This causes TCP to reduce its transmission window in response and often cause long timeouts during loss recovery that keep the connection idle. Thus bit-error losses lead to degraded throughput and end-to-end delay performance.

While TCP adapts well to network congestion, it does not adequately handle the above problems in wireless media. This paper analysed the problems posed by the above challenges, and solved them by using modifications and enhancements to TCP at the sender and receiver.

The DLN protocol is a general framework by which receiver, base station, or other elements in the network can inform the TCP sender of losses that occur for reason other than congestion. When combined with algorithms to distinguish congestion losses from corruption, this framework provides a powerful way by which TCP senders can separate congestion control from loss recovery and recover from non-congestion-related losses without invoking congestion control. Simulation results and mathematical analysis indicate that the DLN protocol can keep the window open big enough without the affect of random wireless error.

It is easy to find the main advantage of DLN compared with other algorithm is its ability to judge the reason of packet loss by using the information contained in the acknowledgement transmitted back from the TCP receiver, thereby avoiding long idle waiting time, this algorithm can significantly improve the end-to-end delay performance in packet transmission.

ACKNOWLEDGEMENTS

I would like to begin by thanking my supervisor, Dr. E. Karthikeyan, Assistant Professor. He has been a wonderful advisor, providing me with support, encouragement, and an endless source of ideas. His breath of knowledge and his enthusiasm for research amazes and inspires me. I thank him for the countless hours he has spent with me, discussing everything from research to career choices, reading my papers, and critiquing my talks. His assistance during the travel of this journal has been absolutely invaluable-my life has been enriched professionally, intellectually, and personally by working with Dr. E. Karthikeyan, this work would have been impossible with his support and assistance. I also thank to anonymous reviewers for their many helpful comments and valuable suggestions.

REFERENCES

[1]   T.Berners-Lee et al, (Aug 1994) "The World Wide Web," Communication of the ACM.

[2]   J.B.Postel and J.Reynolds, (Oct 1985) "File Transfer Protocol(FTP)," Information Sciences Institute, Marina del Rey, CA,. RFC-821.

[3]   J.B.Postel and J.Reynolds, (August 1982) "Simple Mail Transfer Protocol," Information Sciences Institute, Marina del Rey, CA,. RFC-821.

[4]   R.W.Watson and S.A.Mamrak, (May 1987) "Gaining Efficiency in Transport Services by Appropriate Design and Implementation Choices," ACM Transactions on Computer Systems, 5(2):97-120.

[5]   Tom Kelly, (2003) "Scalable TCP: improving performance in highspeed wide area networks," SIGCOMM

[6]   MC.Chan and R.Ramjee, (April 2008) "Improving TCP/IP Performance over Third Generation Wireless Networks", IEEE Transactions on Mobile Computing, 7(4).

[7]   OSI Transport Protocol Specification, 1986. Standard ISO-8703

[8]   W.T.Strayer, B.J.Dempsey, and A.C.Weaver, (1992) "Xtp: The Xpress Transfer Protocol," Addison-Wesley, Reading, MA.

[9]   J. B. Postel, (September 1981) "Transmission Control Protocol," Information Sciences Institute, Marina del Rey, CA.

[10]   D-M. Chiu and R. Jain, "Analysis of the Increase and Decrease Algorithms for Congestion Avoidance in Computer Networks,"

[11]   W.R.Stevens. (November 1994) "TCP/IP Illustrated", Volume 1. Addison-Wesley, Reading, MA.

[12]   P.Karn and C.Partridge, (November 1991) "Improving Round-Trip Time Estimates in Reliable Transport Protocols," ACM Transactions on Computer Systems, 9(4):364-373.

[13]   J.C Hoe, (August 1996) "Improving the Start-up Behavior of a Congestion Control Scheme for TCP", In Proc. ACM SIGCOMM'96.

[14]   M.Mathis and J.Mahadavi, (August 1996) "Forward Acknowledgement Refining TCP Congestion Control," In Proc. ACM SIGCOMM.

[15]   D.Lin and H.Kung, (March 1998) "TCP Fast Recovery Strategies: Analysis and Improvements," In Proc. INFOCOM.

[16]   Ji-Hoon Yun, (2009) "Cross-Layer Explicit Link Status Notification to Improve TCP Performance in Wireless Networks," EURASIP Journal on Wireless Communications and Networking, vol. 2009, Article ID 617818, 15 pages, 2009. doi:10.1155/2009/617818

[17]   W.Ding and A.Jamalipour, (2001) "Delay Performance of the New Explicit Loss Notification TCP Technique for Wireless Networks," GLOBECOM, San Antonio, Texas, Nov.25-29, 2001.

10

LOAD BALANCING BASED APPROACH TO IMPROVE LIFETIME OF WIRELESS SENSOR NETWORK

Dipak Wajgi[1] and Dr. Nileshsingh V. Thakur[2]

[1]Department of Computer Science and Engineering, Shri. Ramdeobaba College of Engineering and Management, Nagpur, India
`wajgi@rediffmail.com`
[2]Department of Computer Science and Engineering, Shri. Ramdeobaba College of Engineering and Management, Nagpur, India
`thakurnisvis@rediffmail.com`

ABSTRACT

In wireless sensor network, clustering is used as an effective technique to achieve scalability, self-organization, power saving, channel access, routing etc.[3]. Lifetime of sensor nodes determines the lifetime of the network and is crucial for the sensing capability.[2]. Clustering is the key technique used to extend the lifetime of a sensor network. Clustering can be used for load balancing to extend the lifetime of a sensor network by reducing energy consumption. Load balancing using clustering can also increase network scalability. Wireless sensor network with the nodes with different energy levels can prolong the network lifetime of the network and also its reliability. In this paper we propose a clustering technique which will balance the load among the cluster by using some backup nodes. The backup high energy and high processing power nodes replace the cluster head after the cluster reaches to its threshold limit. This approach will increase the network lifetime and will provide high throughput.

KEYWORDS

wireless sensor network, clustering , reliability, scalability

1. INTRODUCTION

Information gathering is a fast growing and challenging field in today's world of computing. Sensors provide a cheap and easy solution to these applications especially in the inhospitable and low-maintenance areas where conventional approaches prove to be very costly. Sensors are tiny devices that are capable of gathering physical information like heat, light or motion of an object or environment. Sensors are deployed in an ad-hoc manner in the area of interest to monitor events and gather data about the environment. Networking of these unattended sensors is expected to have significant impact on the efficiency of many military and civil applications, such as combat field surveillance, security and disaster management. Sensors in such systems are typically disposable and expected to last until their energy drains. Therefore, energy is a very scarce resource for such sensor systems and has to be managed wisely in order to extend the life of the sensors for the duration of a particular mission. Typically sensor networks follow the model of a base station or command node, where sensors relay streams of data to the command node either periodically or based on events. The command node can be statically located in the vicinity of the sensors or it can be mobile so that it can move around the sensors and collect data. In either case, the command node cannot be reached efficiently by all the sensors in the system. The nodes that are located far away from the command node will consume more energy to transmit data then other nodes and therefore will die sooner[6].

A wireless sensor network is typically consisting of a potentially large number of resource constrained sensor nodes and few relatively powerful control nodes. Each sensor node has a battery and a low-end processor, a limited amount of memory, and a low power communication module capable of short range wireless communication [3]. As sensor nodes have very limited battery power and they are randomly deployed it is impossible to recharge the dead battery. So the battery power in WSN is considered as scarce resource and should be efficiently used. Sensor node consumes battery in sensing data, receiving data, sending data and processing data [1].

Generally a sensor node does not have sufficient power to send the data or message directly to the base station. Hence, along with sensing the data the sensor node act as a router to propagate the data of its neighbour.

In large sensor network, the sensor nodes can be grouped into small clusters. Each cluster has a cluster head to coordinate the nodes in the cluster. Cluster structure can prolong the lifetime of the sensor network by making the cluster head aggregate data from the nodes in the cluster and send it to the base station. A randomly deployed sensor network requires a cluster formation protocol to partition the network into clusters. The cluster heads should also be selected. There are two approaches used in this process the leader first and the cluster first approach. In the leader first approach the cluster head is selected first and then cluster is formed. In the cluster first approach the cluster is formed first and then the cluster head is selected [3].

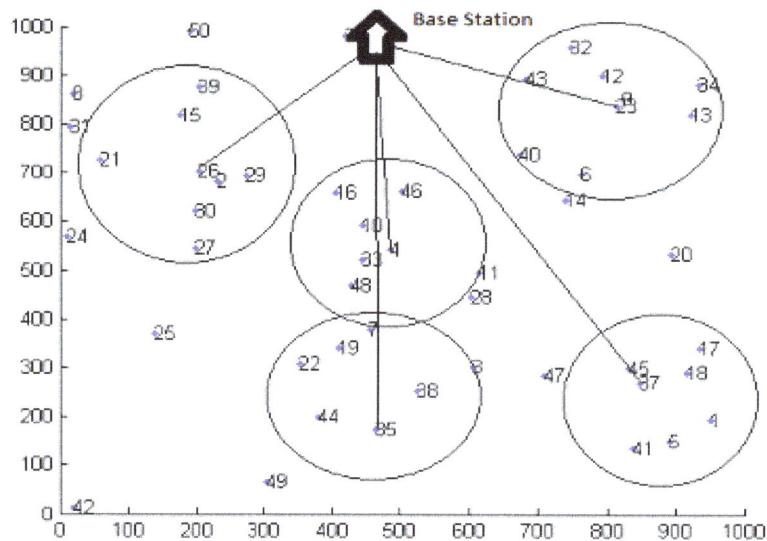

Figure 1. The cluster architecture

Figure1 represent the cluster architecture of the sensor nodes. Initially the sensor nodes are randomly deployed. The nodes are heterogeneous in nature with the different energy levels. The nodes with the higher energy are found in the region and are made cluster head. All the cluster heads defines its range and form the cluster. The cluster member nodes send the sensed data to the cluster heads and cluster head then sends that data to the base station. In most of the cases it is assumed that the cluster head is sending the data to the base station directly.

Clustering has numerous advantages like it reduces the size of the routing table , conserve communication bandwidth, prolong network lifetime, decrease the redundancy of data packets, reduces the rate of energy consumption etc[4].

Generally it is assumed that the nodes in wireless sensor networks are homogeneous, but in reality, homogeneous sensor networks hardly exist. Even homogeneous sensors have different capabilities like different levels of initial energy, depletion rate, etc. In heterogeneous sensor networks, typically, a large number of inexpensive nodes perform sensing, while a few nodes having comparatively more energy perform data filtering, fusion and transport. This leads to the research on heterogeneous networks where two or more types of nodes are considered. Heterogeneity in wireless sensor networks can be used to prolong the life time and reliability of the network. Heterogeneous sensor networks are very popular [4].

1.1 Heterogeneous Model for Wireless Sensor Networks

Different Heterogeneous Model for Wireless Sensor Networks is suggested based on various resources. There are three common types of resource heterogeneity in sensor nodes: computational heterogeneity, link heterogeneity, and energy heterogeneity.

Computational heterogeneity means that the heterogeneous node has a more powerful microprocessor, and more memory, than the normal node. With the powerful computational resources, the heterogeneous nodes can provide complex data processing and longer term storage.

Link heterogeneity means that the heterogeneous node has high bandwidth and long distance network transceiver than the normal node. Link heterogeneity can provide a more reliable data transmission.

Energy heterogeneity means that the heterogeneous node is line powered, or its battery is replaceable.

Among above three types of resource heterogeneity, the most important heterogeneity is the energy heterogeneity because both computational heterogeneity and link heterogeneity consumes more energy resource.

The remainder of the paper is organised as follows. In section 2, review of literature is discussed. In section 3, design philosophy is discussed. In section 4, load balancing approach is proposed and explained. Simulation results are presented in section 5. Section 6 concludes the paper with conclusion and future scope.

2. RELATED WORK

Clustering can be used for balancing the load in the wireless sensor networks[6,8,9,10,12,13,14,18,19,20,21,22,24,25,]. In a cluster based load balancing [6], the maximum transmission power of the nodes is used to become the cluster member. Cluster membership depends on the communication cost. The proposed approach does not consider the backup recovery. A load balanced clustering approach [8], uses comprehensive weight value composed of distance between the head and the member and the residual energy to improve cluster member choice. It also uses optimization threshold value to avoid load imbalance. The algorithm considers load equalization for creating balanced cluster. A multi-hop clustering algorithm for load balancing in wireless sensor networks [9], uses layered approach for intra cluster and inter cluster communication. The algorithm consider homogeneous network. Reconfiguration of cluster head for load balancing in wireless sensor networks [10], increases the network lifetime by fairly distributing the cluster heads. Reconfiguration of the cluster is done based of the number of general nodes in the cluster & the number of cluster heads within the cluster head's transmission range. The algorithm provides effective data aggregation. A novel load balancing scheduling algorithm for wireless sensor networks [12], uses optimal scheduling algorithm for packet forwarding which determines the time slot for sending the packets for the nodes. The algorithm provides uniform packet loss probability for all the nodes. The algorithm uses balanced cost objective function for optimum scheduling. Secure load balancing via hierarchical data aggregation in heterogeneous wireless sensor networks [13], protocol introduces pseudo sink in order to improve data accuracy and bandwidth utilization of wireless sensor network to increase network lifetime. A load balanced algorithm in wireless sensor networks based on pruning mechanism [14], handles the hot point problems which uses pruning mechanism in the cluster to balance the load in the network. Evaluation function in the algorithm is based on pruning mechanism and uses nodes location, residual energy and count of cluster nodes as its parameter to find its cost. In load balancing in energy efficient connected coverage wireless sensor networks [18], the algorithm consider sensing coverage & network connectivity by dividing the sensor network nodes into subsets. It turns on some extra nodes in each subset to ensure network connectivity. The problem with this approach is to find the existence of critical nodes. These nodes may be on all the time and the network will be partitioned if these nodes die. A Threshold Based algorithm for Power aware Load Balancing in Sensor Networks [19], provides possible in-network method for adaptive distributed control of energy consumption. Other methodologies like market based algorithm or game theoretic algorithm can be used. The algorithm assumes complete connectivity. A Load Balanced Clustering Algorithm for Wireless Sensor Networks [20], has Proposed the load balancing algorithm for cluster heads in wireless sensor networks by considering the traffic load as the key parameter. It is assumed that the traffic loads contributed by all the sensor nodes are same, which is the special case of this algorithm. In general case the algorithm is NP hard. It uses centralized approach and assumes that each node is aware of the network. Clustering and Load Balancing in Hybrid Sensor Network with mobile Cluster Nodes[21], has proposed an algorithm that consider the problem of positioning mobile cluster heads and balancing traffic load in hybrid sensor network which consists of static and mobile nodes. It is stated that the location of the cluster head can affect network lifetime significantly. Network load can be balanced and lifetime can be prolonged by moving cluster head to better location. A Load balanced Clustering Algorithm for Heterogeneous Wireless Sensor Networks[22], has Proposed the load balanced group clustering to balance the battery power in wireless sensor network by implementing dynamic route calculation according to the condition of energy distribution in the network. It makes use of heterogeneous energy to realize load balance. Fuzzy Based Approach for Load Balanced Distributing database on Sensor Networks [24], has proposed fuzzy based approach for load balanced distributing database on sensor network that prolong the network lifetime. In this algorithm vertical partitioning algorithm for distributing database on sensors is used. In this approach, first clusters are formed and then distribute partitions on clusters. In an Energy Aware Dynamic Clustering Algorithm for Load Balancing in Wireless sensor network [25], a novel dynamic clustering algorithm for load balanced routing is proposed which is based upon route efficiency. The algorithm uses pattern, traffic load & energy dissipation rate of each node on the route to calculate the node and route efficiency. Energy Efficient Communication Protocol for Wireless Sensor Networks [28], uses low energy adaptive clustering hierarchy that utilizes randomized rotation of local cluster base station to evenly distribute the energy load among sensors of the network. This protocol provides scalability and robustness for dynamic networks. It incorporates data fusion into routing protocol to reduce the amount of information transmitted to the base station. It uses minimum distance from the cluster head to the other nodes as the parameter for the cluster formation. The algorithm is also organized in such a manner that data fusion can be used to reduce the amount of data transmission. The decision of whether a node elevates to cluster head is made dynamically at each interval. The elevation decision is made solely by each node independent of other nodes to minimize overhead in cluster head establishment. This decision is a function of the percentage of optimal on application cluster heads in a network (determined a priory on application), in combination with how often and the last time a given node has been a cluster head in the past. In this algorithm each node calculates the minimum transmission energy to communicate with its cluster head and only transmits with that power level.

Routing can also be used to balance the load in the wireless sensor network [7, 11, 15, 16, 17, 23, 26, 27]. Multipath Routing Algorithm for wireless sensor network [7], finds the node disjoint routing path similar to mazing search. It reduces the energy consumption and congestion. It introduces Multi path selection strategy to balance the load in the network. The bandwidth is utilized efficiently for transmitting audio and video packets. Energy Aware Intra Cluster Routing for wireless sensor network [11], performs adaptive routing where the distance from base station is taken into consideration. Multihop Routing Based on Optimization of the number of cluster heads in wireless sensor network [15], has proposed the equation for the number of packets to send and relay to calculate the energy consumption of sensor nodes. Change in the number of cluster heads affects the consumed energy of sensor nodes. Overlapping Multihop Clustering for wireless sensor network [16], has proposed Randomized, distributed multihop clustering protocol for solving overlapping clustering problems. It considers average overlapping degree. It is scalable. Cluster formation terminates in constant time regardless for network size. A Novel Cluster Based Routing Protocol with Extending Lifetime for Wireless Sensor Networks [17], has self configuration and hierarchical routing properties. It construct cluster based on radio radius and number of cluster members. Clusters in the network are equally distributed. Sensor nodes perform voting for selecting cluster head. An Energy Aware Cluster Based Routing Algorithm for Wireless Sensor Networks [23], has proposed the algorithm for wireless sensor network to maximize network's lifetime. It selects some nodes as cluster heads to construct voronoi diagram and rotate the cluster head to balance the load in each other. A cluster Based Energy Efficient Location Routing Protocol in Wireless Sensor Networks [26], uses hierarchical structured method, multihop and location based nodes. It works well for small networks. Cluster head is selected based on residual energy and minimum distance from the base station. In a Novel Cluster Based Routing Protocol in Wireless Sensor Networks [27], a cluster based routing protocol for prolonging the sensor network lifetime is proposed. The algorithm achieves a good performance in terms of lifetime by balancing the energy load among all the nodes. Cluster Based Routing Protocol achieves a good performance in terms of lifetime by balancing the energy load among all the nodes. In this protocol first the clusters are formed then the spanning tree is constructed for sending aggregated data to the base station which can better handle the heterogeneous energy capacities.

3. DESIGN PHILOSOPHY

Wireless Sensor Networks present vast challenges in terms of implementation. Clustering algorithms play a vital role in achieving the targeted design goals for a given implementation. There are several key attributes that must be considered in wireless sensor networks [5].

Selection of Cluster heads and Clusters: The clustering concept offers tremendous benefits for wireless sensor networks. However when designing for a particular application, designers must carefully examine the formation of clusters in the network. Depending on the application, certain requirements for the number of nodes in a cluster or its physical size may play an important role in its operation. This prerequisite may have an impact on how cluster heads are selected.

Data Aggregation: One major advantage of wireless sensor networks is the ability for data aggregation to occur in the network. In a densely populated network there are often multiple nodes sensing similar information. Data aggregation allows the differentiation between sensed data and useful data. Network processing makes this process possible and now it is fundamental in many sensor network schemes, as the power required for processing tasks is substantially less than communication tasks. As such, the amount of data transferred in network should be minimized. Many clustering schemes provide data aggregation capabilities, and as such, the requirement for data aggregation should be carefully considered when selecting a clustering approach.

Repair Mechanisms: Due to the nature of Wireless Sensor Networks, they are often prone to node mobility, node death and interference. All of these situations can result in link failure. When looking at clustering schemes, it is important to look at the mechanisms in place for link recovery and reliable data communication.

4. PROPOSED APPROACH

The proposed approach assumes heterogeneous network with the sensor nodes having different energy levels and processing power. Some high computing nodes are deployed nearby each other. All the nodes with high initial energy level and processing power are selected. Some nodes from the set are selected as cluster head (CH) according to their location. Each CH defines its communication range in terms of power level to form cluster. Some nodes with comparable energy and processing power in the CH range are asked to go to sleep and information about those nodes is maintained with the CH. Each CH sends a hello request message to all the nodes within its communication range to become the cluster member. This process will be repeated for all the CH. All the cluster members will send the sensed data to the CH. The CH will send the aggregated data to the Base Station directly or by using some intermediate CH.

When the energy level of the CH will reach to the threshold value TL, the CH will activate one of the sleeping nodes and will make it CH. This information about the new CH will be sent to all the cluster member and other CH also. The old CH will become the general sensor node.

4.1 ALGORITHM

The algorithm is divided into four phases
1. Initialization Phase
1.1 Select the CH according to the capabilities of the nodes.
1.2 Select the desired number of CH according to their location.
1.3 Define the range of CH.
1.4 CH sends membership request message to all the nodes in its range and request to reply with their current energy status.
1.5 The nodes with high residual energy and processing power will be identified and they are made to sleep. They become the backup nodes.
1.6 The nodes which are not in the range of cluster head, will try to join the cluster by sending the message to the nearest cluster member.
2. Steady State Phase
2.1 The cluster members sends the sensed data to the CH in the allotted time using TDMA schedule.
2.2 The non cluster members will send the sensed data to the cluster head through the intermediate cluster member.
3. Final Phase
3.1 CH will aggregate the data from all the nodes in its cluster.
3.2 CH will transmit the data to the base station.
4. Cluster Reconfiguration Phase
4.1 If the CH residual energy reaches to the threshold value, the CH will activate the backup node.

4.2 The CH will relegate its responsibility to the backup node and will make the node the cluster head.

4.3 The CH will transmit the new CH information to all other nodes in the cluster.

4.4 The CH will transmit the new CH information to all other CH also.

4.5 The old CH will become the general node.

5. EXPERIMENTAL SETUP AND RESULTS

The sensor nodes are assumed to be distributed randomly.

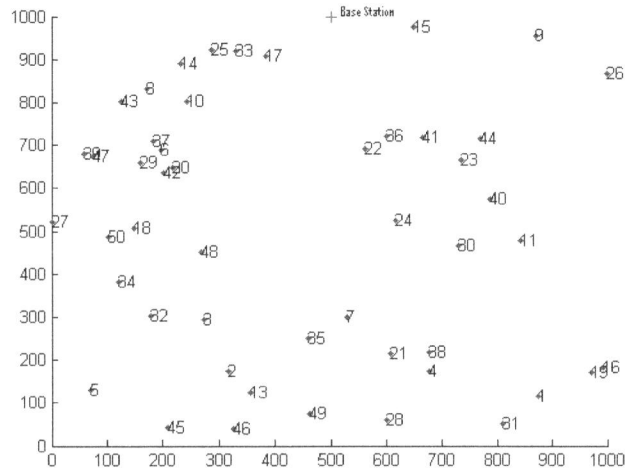

Figure 2. The random distribution of the sensor nodes.

Figure 2 represent the random distribution of the sensor nodes in MATLAB. The flat area is considered for the distribution of the sensor nodes. The position of the base station is also fixed. In this random distribution, the areas are found where the density of the sensor nodes is more. Accordingly the nodes with the higher energy and high processing power will be deployed. While deploying the high energy and high processing power nodes the care is taken that the backup nodes are also deployed nearby the high energy and high processing power nodes. Here in this case, consider these nodes are 37 and 6, 2 and 13, 4 and 38 and 23 and 40. We are creating the pairs of the nodes so that if one node fails or reaches to the threshold energy value the backup node will take the responsibility of that node.

5.1 CLUSTERING PROCESS

Cluster formation process is shown in the Figure 3. The high energy and high processing power nodes which are the cluster heads defines their range of communication in terms of distance. In the above figure the nodes 37, 2,4 and 23 are acting as cluster heads. Then they send the membership request message to the sensor nodes in their communication range. The sensor nodes are requested to send the acceptance message along with their energy status to the requesting cluster head. After receiving the acceptance message and the energy information, the cluster head finds the node with the energy and power comparable to it. If found, it allows the node to go to sleep. This node acts as a backup node for the cluster head. In the above figure the node 6 is acting as a backup nodes for node 37, node 13 is acting as a backup nodes for node 2, node 38 is acting as a backup nodes for node4, node 40 is acting as a backup nodes for node

23. In the sleeping mode also the sensor node is dissipating the energy but with very less rate. All the cluster heads forms their clusters along with the backup nodes.

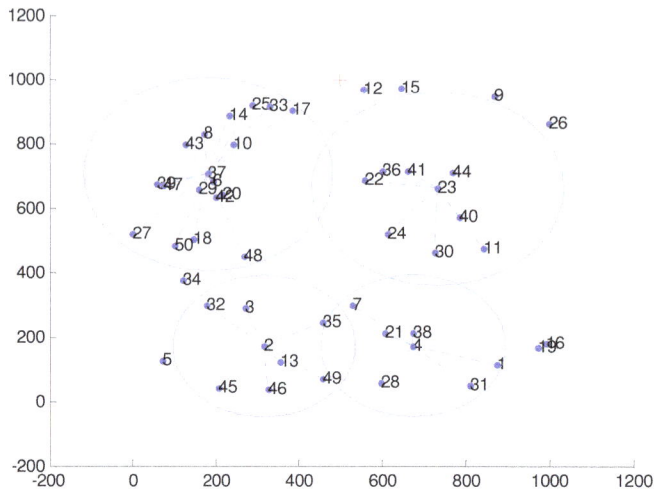

Figure 3. Cluster formation

There are some nodes which are not the member of any cluster. These nodes search for the minimum distance node which is the member of any one of the cluster. In the figure 4, node 12 is attached to the cluster of cluster head 37 through the node 17 and node 15 is attached to the cluster of cluster head 23 through node 41. Like this all the remaining nodes are attached to the cluster through the intermediate node which is the cluster member of the particular cluster.

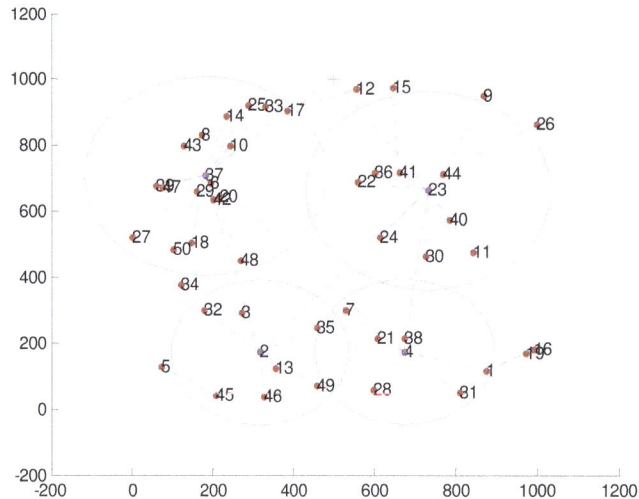

Figure 4. Connection among the cluster heads and non cluster member nodes.

5.2 RECONFIGURATION OF CLUSTERS

The cluster reconfiguration process is shown in Figure 5. After some time period which is modelled as rounds, the energy level of the cluster head reaches to the threshold value. This threshold value is already set which is the indicator that the cluster head can no longer handle the responsibility of the head and should hand over the responsibility to the backup node. The cluster head awake the backup node which is in the sleep node and ask it to send the hello message to the nodes in the range. At the same time it sends the message to all the member nodes informing about the new cluster head. If any member node, not getting the hello message from the new cluster head, it means that the node is not within the range of the new cluster head and should find the new cluster. This process is followed by all the cluster heads. In this process the membership of the nodes can be changed i.e. the node which is the member of one cluster may become the member of another cluster after reconfiguration and the nodes which is not the direct member of any cluster may become the direct member of the cluster.

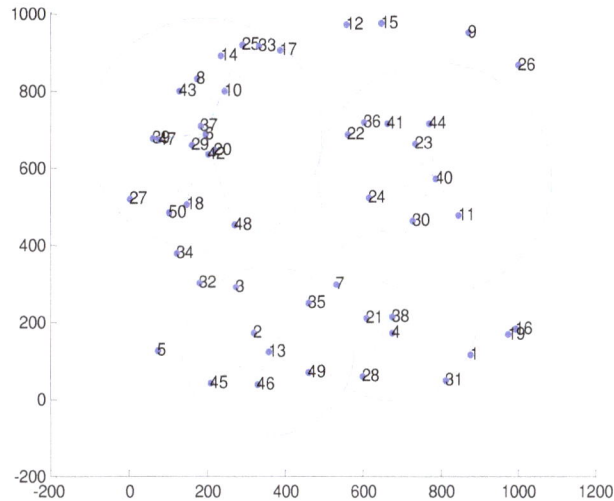

Figure 5. Reconfiguration of cluster

After completing this reconfiguration phase, some nodes, which are not the members of any of the cluster, try to find the nearest cluster member node. Again, the same criterion of finding the minimum distance cluster member is applied to find the appropriate cluster. The Figure 6 shows the connection among the new cluster heads and the remaining nodes. So these remaining nodes are indirectly attached to the new clusters.

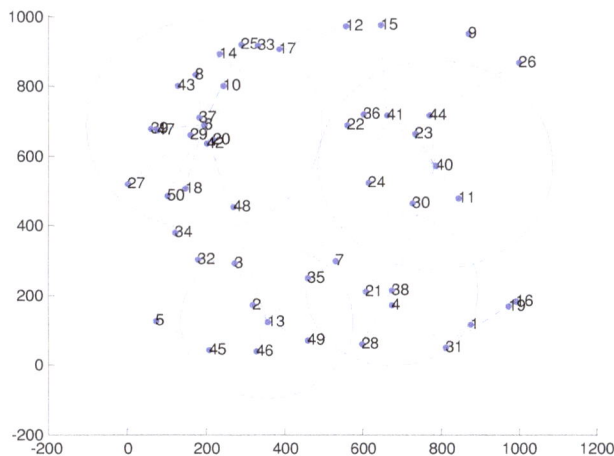

Figure 6. Connection among the new cluster heads and non cluster member nodes.

5.3 PERFORMANCE GRAPH

The performance of the proposed algorithm is plotted in the graph showing the number of the nodes against the number of rounds in the Figure 7. The round starts with the initialization phase of creating the clusters. After the clusters are formed the sensor nodes are aggregating the data and sending it to the cluster heads. Before this the base station is giving some instructions to the cluster heads for collecting some information in the particular area. The cluster heads pass on this information to the cluster member nodes. In turn, the member nodes collect the information and send it to the cluster head and then cluster head sends the collected information to the base station for the further processing. The energy of the sensor node is consumed while sensing the data, performing some operations on data. But most of the energy of the sensor nodes is consumed during communication. The dissipation of the energy during communication is depending on the distance between the sensor nodes. If the distance is longer then more energy gets dissipated and if the distance is small then less energy is dissipated.

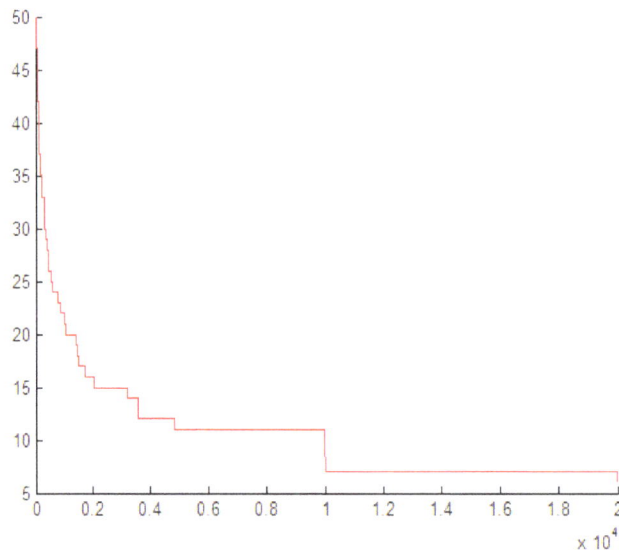

Figure 7. Performance graph showing the number of nodes on y-axis and the rounds on the x-axis.

If all the sensor nodes are sending the data to the base station directly, then the lifetime of the sensor network gets reduced since the base station is located far away from the senor nodes and the energy dissipated for this long haul communication is very much high. This is one of the reasons behind going for clustering techniques. The sensor node communication is limited within the cluster. So the energy dissipated will be less. The cluster heads are responsible for communication with the base station. Being the high energy and high processing power nodes, they perform the long haul communication and thus prolong the network lifetime.

If all the cluster heads are sending the data to the base station then some of the cluster heads may die sooner since they are at long distance. So some routing mechanism is also employed to prolong the network lifetime. The long distance cluster heads send the data to the base station via some intermediate cluster heads.

If all the long distance cluster heads are sending the data to the cluster heads which are close to the base station, then there are chances that the cluster heads close to the base station will die sooner, since along with the data aggregation and sending their own data they have to aggregate the data from the other cluster heads and send it to the base station. To avoid this, routing in used where the distant cluster heads will not send the data to the same cluster head near the base station in every round but they will send the data every alternate round.

The nodes which are not the direct member of the cluster send the data to the member node of the cluster and in turn the member node along with its own data sends it to the cluster head. Energy dissipation of the member nodes and non-member nodes depends on the distance from the cluster head. If the distance is more, then more energy is dissipated otherwise less energy is dissipated.

In the Figure 7, the graph is plotted for the number of live nodes against the number of rounds. The time from the initialization phase to the first node death defines the network lifetime. If the results are compared with the original LEACH algorithm, the first clustering algorithm, with the same experimental setup, then the results are better. Even with the first node death, unless and until all the cluster head die the network is performing its job.

6. CONCLUSION AND FUTURE SCOPE

In this paper, an approach for load balancing in the wireless sensor network is proposed. Algorithms for cluster head selection, cluster formation, intra cluster communication and inter cluster communication in wireless sensor network are proposed. The performance of the algorithm is compared with the original LEACH algorithm with respect to the number of rounds and the dead nodes using the parameter like energy dissipation in each round per node. The results demonstrate that the proposed approach is effective in prolonging the network lifetime.
In future, experiments are planned to be extended for parameters and scenarios like coverage, fault tolerance, impact of aggregation and mobility of nodes.

The proposed approach is implemented by using MATLAB. In future the tools like NS-2 and OMNET++ can be used for implementation.

References

[1]   Babar Nazir, Halabi Hasbullah," Energy Balanced Clustering in Wireless Sensor Network", 978-1-4244-6716-7/10, 2010 IEEE.

[2]   Neeta Trivedi, G. Elangovan, S.S. Iyengar, N. Balakrishnan, "A Message-Efficient, Distributed Clustering Algorithm for Wireless Sensor and Actor Networks", 1-4244-0567-x/06, 2006 IEEE.

[3]   Kun Sun, Pai Peng Peng Ning, "Secure Distributed Cluster Formation in Wireless Sensor Networks", Proceedings of the 22nd Annual Computer Security Applications Conference (ACSAC'06), pages 131--140, December 2006.

[4]   Vivek Katiyar, Narottam Chand, Surender Soni, "Clustering Algorithms for Heterogeneous Wireless Sensor Network: A Survey", IJAER, Volume1, No 2, 2010, ISSN-0976-4259

[5]   D. J. Dechene, A. El Jardali, M. Luccini, A. Sauer, "A Survey of Clustering Algorithms for wireless Sensor Networks", Information and Automation for Sustainability, 2008. ICIAFS 2008. 4th International Conference , Publication Year: 2008 , Page(s): 295 - 300

[6]   G. Gupta, M. Younis, "Load-Balanced Clustering in Wireless Sensor Networks", Pproceedings of International Conference on Communications (ICC 2003), Page(s): 1848 - 1852 , Vol.3 , Anchorage, Alaska, May2003

[7]   M. Xie, Y. Gu, "Multipath Routing Algorithm for Wireless Multimedia Sensor Networks within Expected Network Lifetime", 2010 International Conference on Communication and Mobile Computing 12-14 April 2010,Volume:3 Page(s): 284 – 287, IEEE.

[8]   H. Zhang, L. Li, X. Yan, X. Li, "A Load Balancing Clustering Algorithm of WSN for Data Gathering", 978-1-4577-0536-6/11, 2011, IEEE.

[9]   N. Israr & I. Awan, "Multi-hop clustering Algo. For Load Balancing in WSN", 2006 International Journal of Simulation Vol. 8 No. 1 ISSN 1473-804x

[10]   N. Kim, J. Heo, H. S. Kim & W. H. Kwon, "Reconfiguration of Cluster head for Load Balancing in Wireless Sensor Networks", Science Direct Computer Communications 31 (2008) 153–159.

[11]   A. Akhtar, A. A. Minhas & S. Jabbaar, "Energy Aware Intra Cluster Routing for Wireless Sensor Networks", International Journal of Hybrid Information Technology, Vol.3, No.1, January, 2010.

[12]   E. Laszlo, K. Tornai, G. Treplan & J. Levendovszky, "Novel Load Balancing Scheduling Algo. For Wireless Sensor Networks", CTRQ 2011 : The Fourth International Conference on Communication Theory, Reliability, and Quality of Service IARIA, 2011.

[13]   S. Ozdemir, "Secure Load Balancing via Hierarchical Data Aggregation in Heterogeneous Sensor Networks", Journal of Information Science And Engineering 25, 1691-1705 (2009).

[14]   Y. Zhang, Z. Zheng, Y. Jin, X. Wang," Load Balanced Algorithm In Wireless Sensor Networks Based on Pruning Mechanism", IEEE transaction, 978-0-7695-3522-7/09.

[15]   C. S. Nam, Y. S. Han & D. R. Shin, Multihop Routing Based on Optimization of the number of cluster heads in Wireless Sensor Networks, Sensors 2011, 11, 2875-2884; doi:10.3390/s110302875.

[16]   M.A. Youssef, A. Youssef & M.F.Younis, "Overlapping Multihop Clustering for WSN", IEEE Transactions On Parallel And Distributed Systems, Vol. 20, No. 12, December 2009

[17]   I.J.Lotf, M.N.Bonab & S. Khorsandi, "A Novel Cluster Based Routing Protocol with Extending Lifetime for Wireless Sensor Networks", 2008, IEEE, 978-1-4244-1980-7/08

[18]   M. Mahdari, M. Ismail, K. Jumari, "Load Balancing in Energy Efficient Connected Coverage Wireless Sensor Networks", IEEE Transaction, Volume: 02 pp. 448-452, 2009

[19]   C. M. Canci, V. Trifa & A. Martinoli, "Threshold Based Algo. For Power Aware Load Balancing in Sensor Networks", IEEE Transaction, 0-7803-8916-6/05, 2005.

[20]   C.P.Low, C. Fang, J. Mee, Ng & Y.H. Hang, "Load Balanced Clustering Algorithm for Wireless Sensor Networks", IEEE Communications Society, 1-4244-0353-7/07, 2007.

[21]   Ming Ma & Y. Yang, "Clustering & Load Balancing in Hybrid Sensor Network with mobile Cluster Heads", Qshine '06', proceedings of the 3rd international conference on quality of service in heterogeneous wired/ wireless Networks, 2006.

[22]   Y. Deng, Y. Hu, "A Load balanced Clustering Algorithm for Heterogeneous Wireless Sensor Networks", 2010, IEEE, pp. 1-4, ISBN: 978-1-4244-7159-1.

[23]   J. H. Chang, "An Energy Aware Cluster Based Routing Algorithm for Wireless Sensor Networks", Journal of Information Science & Engineering, 26, 2159-2171, 2007.

[24]   M. Zeynali, L.M.Khanli, A. Mollanejad, "Fuzzy Based Approach for Load Balanced Distributing database on Sensor Networks", International Journal of Future Generation Communication & Networking, Vol. 3, No. 2, June 2010

[25]   M. Iqbal, I. Gondal & L. Dooley, "An Energy Aware Dynamic Clustering Algorithm for Load Balancing in Wireless Sensor Networks", Journal Of Communications, Vol. 1, No. 3, June 2006

[26]   Nurhayati, S.H.Choi & K.O. Lee, "A cluster Based Energy Efficient Location Routing Protocol in Wireless Sensor Networks", International Journal of Computers And Communications, Volume 5, Issue 2, 2011

[27]   Bager Zarei1, Mohammad Zeynali   and Vahid Majid Nezhad, "Novel Cluster Based Routing Protocol in Wireless Sensor Networks", IJCSI International Journal of Computer Science Issues, Vol. 7, Issue 4, No 1, July 2010.

[28]   W. R. Heinzelman, A. Chandrakasan,   H. Balakrishnan, "Energy-Efficient Communication Protocol for Wireless Microsensor Networks", Proceedings of the 33rd Annual Hawaill International Conference on System Sciences, Jan 4-7, 2000, Maui. USA. Los Alamitos, CA, USA: IEEE Computer Society, 2000: 223

11

An ANN Based Call Handoff Management Scheme for Mobile Cellular Network

P. P. Bhattacharya[1], Ananya Sarkar[2], IndranilSarkar[3], Subhajit Chatterjee[4]

[1]Department of ECE, Faculty of Engineering and Technology, Mody Institute of Technology & Science (Deemed University), Rajasthan, India
[2]Department of ECE,College of Engineering and Management Kolaghat, West Bengal, India
[3]Department of ECE,Sobhasaria Group of Institutions, Rajasthan, India
[4]Deparment of ECE, Swami Vivekananda Institute of Science and Technology, Barruipur, West Bengal

Abstract

Handoff decisions are usually signal strength based because of simplicity and effectiveness. Apart from the conventional techniques, such as threshold and hysteresis based schemes, recently many artificial intelligent techniques such as Fuzzy Logic, Artificial Neural Network (ANN) etc. are also used for taking handoff decision. In this paper, an Artificial Neural Network based handoff algorithm is proposed and it's performance is studied. We have used ANNhere for taking fast and accurate handoff decision. In our proposed handoff algorithm, Backpropagation Neural Network model is used.The advantages of Backpropagation method are its simplicity and reasonable speed. The algorithm is designed, tested and found to give optimum results.

Keywords

Handoff; Backpropagation; Artificial Neural Network; Received Signal Strengths; Traffic Intensities.

1. INTRODUCTION

In mobile cellular communication, maintaining continuous communication when the user migrates from one cell to another is done by changing the controlling base station – a process called call Handoff. Handoff involves measurement, decision and execution. In present generation mobile cellular systems, Mobile Station (MS) estimates the signal strengths from each base station and the value of the received signal level is generally affected by three parameters : path loss, shadow fading and small scale fading. Small scale fading has much shorter correlation distance and averaged out over the time scale under consideration [1] and also anti-multipath fading techniques are available now-a-days [2, 3]. Hence, in a system with anti-multipath technique the effect of small scale fading is not normally considered. But in the present work, multipath fading is considered for considering practical scenarios.

In practice, the low speed mobiles may stop after the handoff execution resulting unnecessary handoff. Similarly, the high speed mobiles may move well into the next cell before the handoff execution resulting call termination. Moreover, the signal strength from base station decreases as exp $(-\gamma d)$ where d is the distance of the mobile station from base station and γ is the path loss exponent. In uniform propagation environment, γ can be taken as constant. But in real environment γ may have different values at different places varying from 2 to 6. So, an algorithm based on path loss exponent and user velocity is essential.

As discussed above, handoff characteristics are user velocity dependent. The effect of mobile velocity on handoff performance has been studied by many workers [4,5,6]. Performance metrics such as probability of handoff, average number of handoff, call blocking probability and call completion probability change significantly as user velocity changes. The traffic density in an average urban area generally follows normal distribution [7]. In our country, the average speed in four metro cities e.g, Delhi, Mumbai, Chennai and Kolkata were found to be 30 Km/hr, 25 Km/hr, 25 Km/hr and 22 Km/hr respectively [8]. Due to the sensitivity of handoff performance to path loss exponent, as discussed in the previous section, a variable hysteresis scheme is already proposed [9] where the hysteresis margin is determined as a function of path loss exponent.

In our work, signal strength from the serving and target base stations and traffic intensities of the serving and target base stations are considered. A three layer ANN model [10] is chosen in the design. Signal strengths from the serving and target base stations are estimated using least square estimation method incorporating Rayleigh fading [11]. A Threshold and hysteresis margin based scheme is chosen where handoff decision is taken only when the signal strength from the current base falls below some threshold value and also the signal strength difference between the current and the serving base station is higher than the hysteresis margin so as to avoid ping – pong effect. In the proposed handoff scheme different signal strengths and traffic intensities are considered to find out the position of handoff. Simulation is carried out using C++ language.

2. BACK PROPAGATION NEURAL NETWORK

An ANN which is an information-processing paradigm is configured for a particular application through a learning process. In our proposed algorithm, Backpropagation Neural Network is used which is an iterative method that starts with the last layer and moves backward through the layers until the first layer is reached. In this method, the outputs and the errors in outputs are calculated and the weights on the output units are altered. Then the errors in the hidden nodes are calculated and the weights in the hidden nodes are altered. The Backpropagation algorithm changes the weights to minimize the errors. The Backpropagation (**BP**) structure shown in Fig.1 consists of three groups, or layers, of units: a layer of "**input**" units is connected to a layer of "**hidden**" units, which is connected to a layer of **"output"** units. The activity of the input units represents the raw information that is fed into the network. The activity of each hidden unit is determined by both the activities of the input units and the connectivity weights between the input and the hidden units. The behavior of the output units depends on the activity of the hidden units and the weights between the hidden and output units.

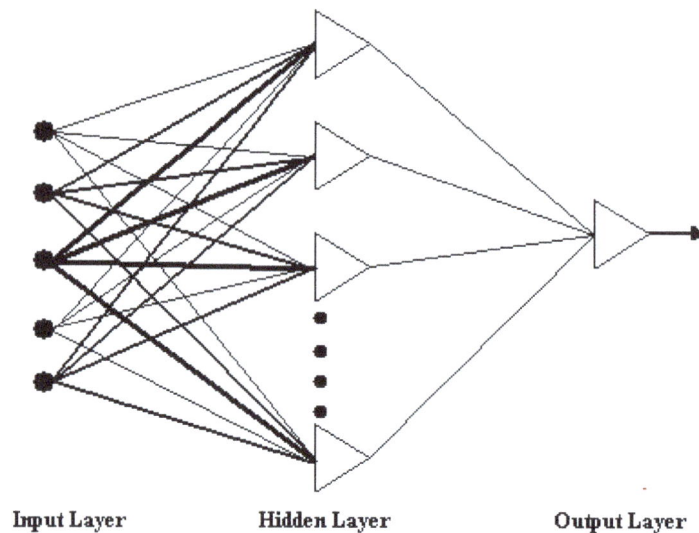

Fig. 1 Backpropagation Neural Network

The number of nodes used in the hidden layer is 20. This number is found after training the network and the errors were found to converge using the value [12]. The output node is a linearly weighted sum of the hidden unit outputs. The outputs decide whether the system needs a handoff or not. If output is -1, no handoff decision will be taken. If output is +1, then handoff decision will be taken. This simple type of network is interesting because the hidden units are free to construct their own representations of the input. The weights between the input and hidden units determine when each hidden unit is active, and so by modifying these weights, a hidden unit can choose what it represents. The advantages of back propagation method are its simplicity and reasonable speed.

Selection of a good activation function is very important because it should be symmetric, and the neural network should be trained to a value that is lower than the limits of the function. One good selection for the activation function is the hyperbolic tangent, or **F(y) = tanh(y)**, because it is completely symmetric, as shown in Fig 2.

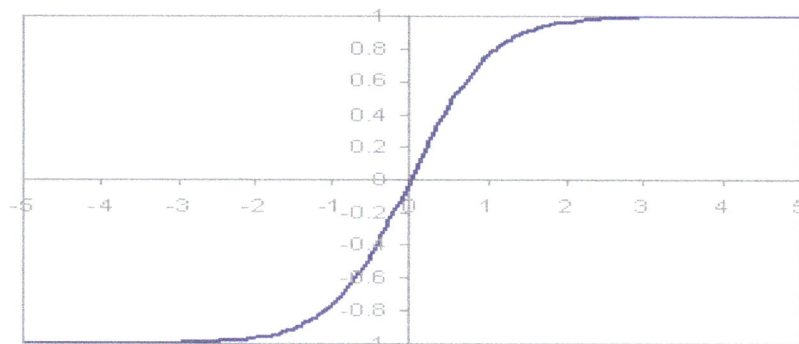

Fig. 2 Hyperbolic tangent function

Another reason for chosing it is that it's easy to obtain its derivative and also the value of derivative can be expressed in terms of the output value (i.e., as opposed to the input value). In our work, this hyperbolic tangent function is chosen.

3. PROPOSED HANDOFF ALGORITHM

Two base stations are considered in our paper and the cell radius is assumed to be 500 meter. Fig 3 is the flow chart illustrating the proposed handoff algorithm. Signal strengths of the serving and target base stations are monitored. When the received signal strength from the serving BS less than the threshold value and the received signal strength (**RSS**) from the serving BS is lower than the target BS by hysteresis margin, then a handoff is done to continue the call in progress. Otherwise no Handoff decision will be taken. Then Artificial Neural Network is used to take the handoff decision depending on both **RSS**s and **TI**s [1] of the serving and target BSs. If output of the neural network is +1 then handoff decision should be taken, where as for –1 no handoff will be taken. The threshold value and the hysteresis margin are chosen to be -85 dBm and -5 dBm respectively.

The inputs to the neural network are listed below -

1. x1 is the signal strength of mobile received from the serving BS.
2. x2 is the signal strength of mobile received from the target BS.
3. x3 is the traffic intensity (TI) of the serving BS.
4. x4is the TI of the target BS.
5. x5 is the bias.

The received signal strength (**RSS**) is considered as:

Low (L): $RSS \leq -85$ dBm and **High (H)**: $RSS > -85$ dBm.

The Traffic Intensity is considered as follow:

Low (L): $TI < 0.66$ Erlangs/Channel,

Medium (M): $0.66 \geq TI \geq 0.76$ Erlangs/ Channel,

High (H): $TI > 0.76$ Erlangs/Channel,

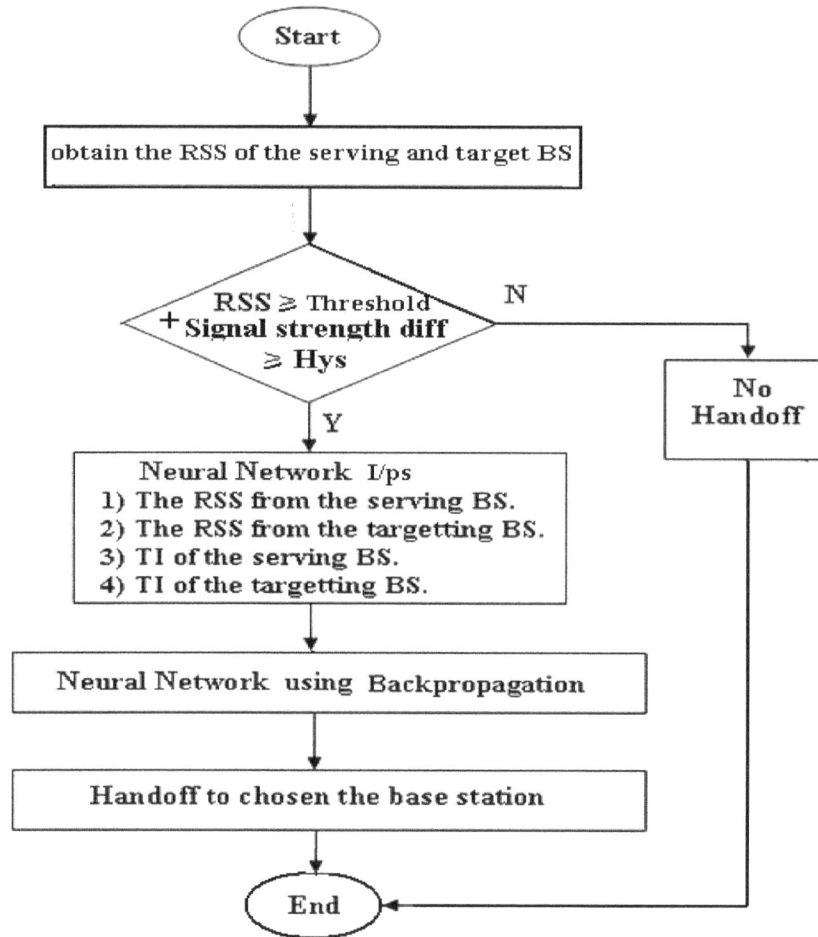

Fig.3 Neural Network based Handoff Algorithm

In this paper four different cases are considered as mentioned in Table 1, such as:

1) Both the **RSS**s from the serving and target **BS**s are low.

2) The **RSS** from the serving **BS** is low while the **RSS** of the target **BS** is high.

3) Both the **RSS**s from the serving and target **BS**s are high.

4) The **RSS** from the serving **BS** is high while the **RSS** from the target **BS** is low.

In each case handoff decisions (**HO**: handoff or **NOHO**: no-handoff) will be taken depending on the different levels of traffic intensities. Let us consider that **RSS** from the serving cell is low and the **RSS** of target cell is high and their traffic intensities are low and high respectively. It is observed that neural network decides not to initiate handoff, as it is desired.

RSS from the target BS : LOW	RSS from the serving BS : LOW		
	TI : S / TI : T	LOW	MEDIUM	HIGH
	LOW	NOHO	HO	HO
	MEDIUM	NOHO	NOHO	HO
	HIGH	NOHO	NOHO	NOHO

RSS from the target BS : HIGH	RSS from the serving BS : HIGH		
	TI : S / TI : T	LOW	MEDIUM	HIGH
	LOW	NOHO	HO	HO
	MEDIUM	NOHO	NOHO	HO
	HIGH	NOHO	NOHO	NOHO

RSS from the target BS : HIGH	RSS from the serving BS : LOW		
	TI : S / TI : T	LOW	MEDIUM	HIGH
	LOW	HO	HO	HO
	MEDIUM	HO	HO	HO
	HIGH	NOHO	NOHO	HO

RSS from the target BS : LOW	RSS from the serving BS : HIGH		
	TI : S / TI : T	LOW	MEDIUM	HIGH
	LOW	NOHO	NOHO	HO
	MEDIUM	NOHO	NOHO	HO
	HIGH	NOHO	NOHO	NOHO

Table 1. Handoff decision

4. RESULTS AND DISCUSSIONS

The estimated signal strengths [5] from serving and target BS are shown in Fig.4. It is observed that received signal strengthsfluctuatein a random manner in a Rayleigh fading environment. For different values of hysteresis margin and threshold value, the possibilities of handoff against distance from serving base station are calculated for different traffic intensities (Fig.5, 6, 7). It is observed from the Fig.5,that for L/L or M/M or H/H or L/M **(Low :L, Medium :M, High :H)** traffic intensity combinations the distance at which handoff decision is taken remain same. Again for H/L or H/M or M/L traffic intensity combinations the distance at which handoff decision is taken remain same as shown in Fig.6. While for L/H or M/H traffic intensity combinations no handoff will be initiated as shown in Fig.7. The results show quick response and minimum fluctuations in handoff decision. Moreover, average numbers of handoffs versus different hysteresis margins and threshold values are calculated for different sets of traffic intensities (Fig.8.). It is observed from the Fig.8.(a), that for L/L or M/M or H/H or L/M traffic intensity combinations the average numbers of handoffs is same for different hysteresis margins and different threshold values. Again for H/L or H/M or M/L traffic intensity combinations the average numbers of handoffs is same for different hysteresis margins and different threshold values as shown in Fig.8.(b). While for L/H or M/H traffic intensity combinations no handoff will be initiated as shown in Fig.8(c). Thus the algorithm works well under all possible situations.

Hysteresis = -5 dBm and Threshold = - 85 dBm
(Low: L, Medium: M, High: H)

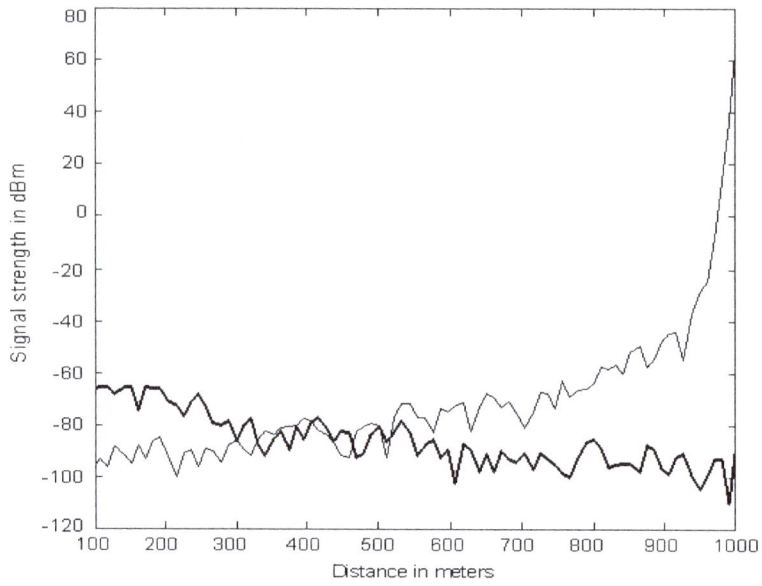

Fig.4 Received signal strengths ofserving and target base stations

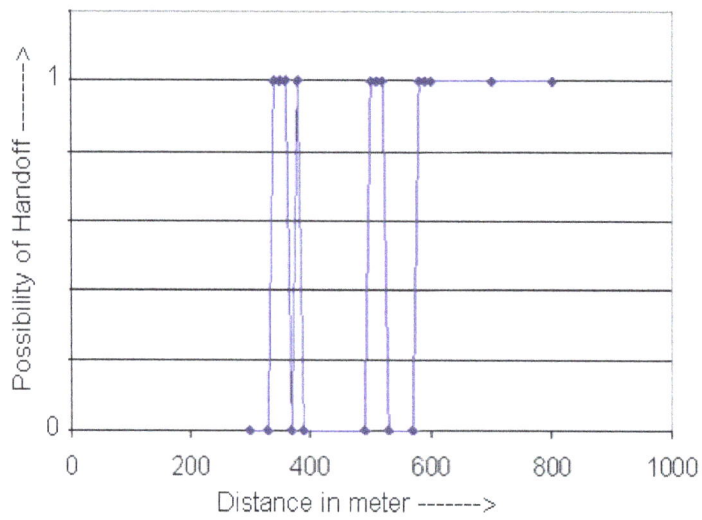

Fig. 5 For L/L or M/M or H/H or L/M   Traffic Intensities

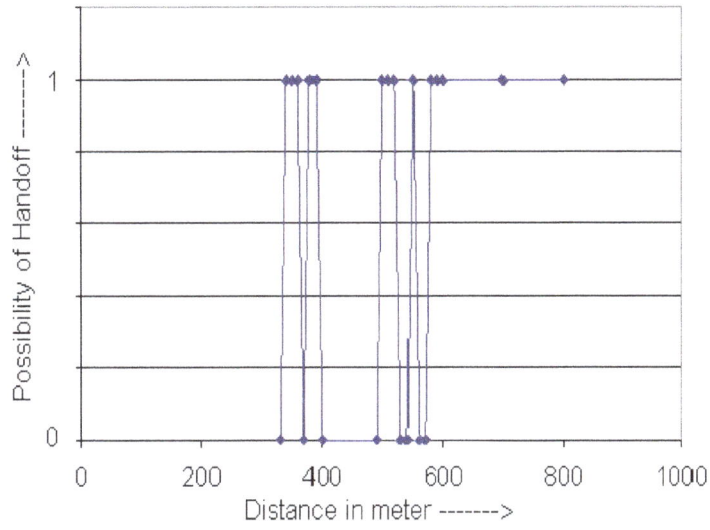

Fig. 6ForH/L or H/M or M/L Traffic Intensities

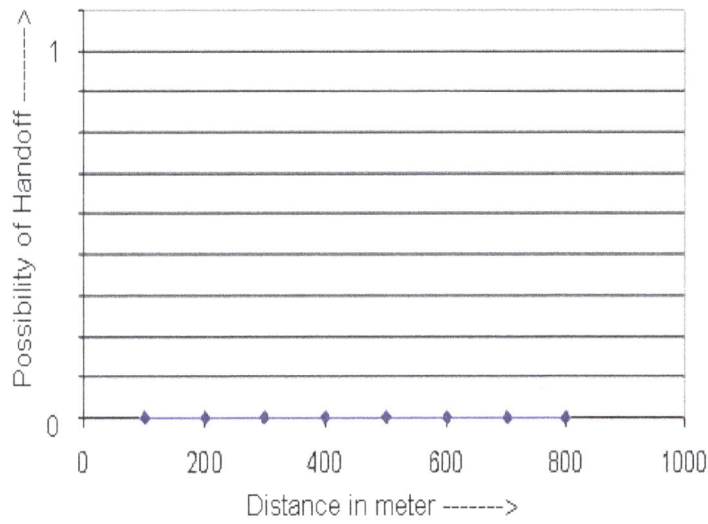

Fig. 7For L/H or M/H Traffic Intensities

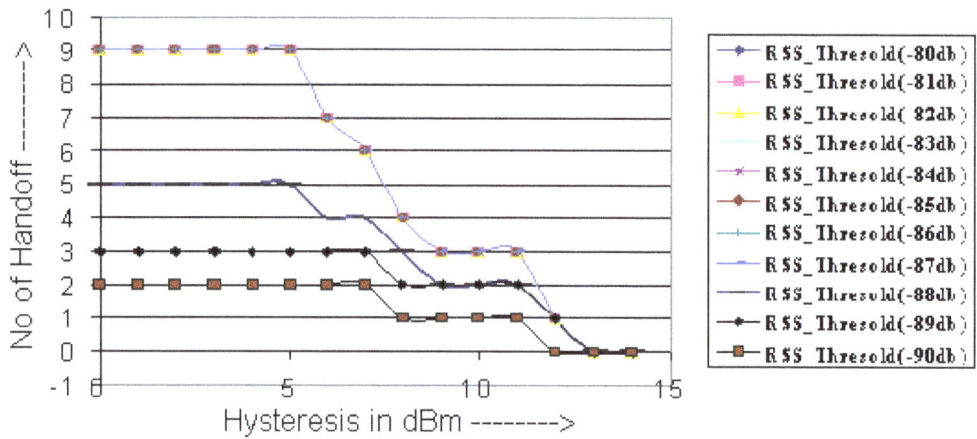

(a)

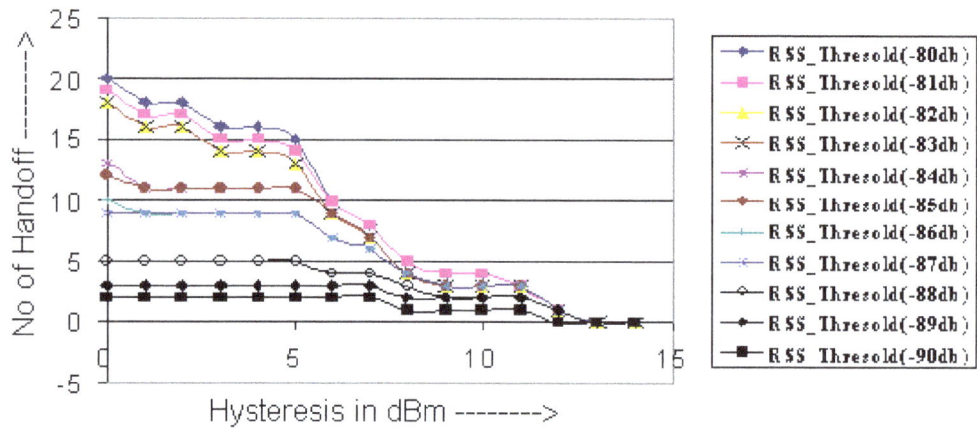

(b)

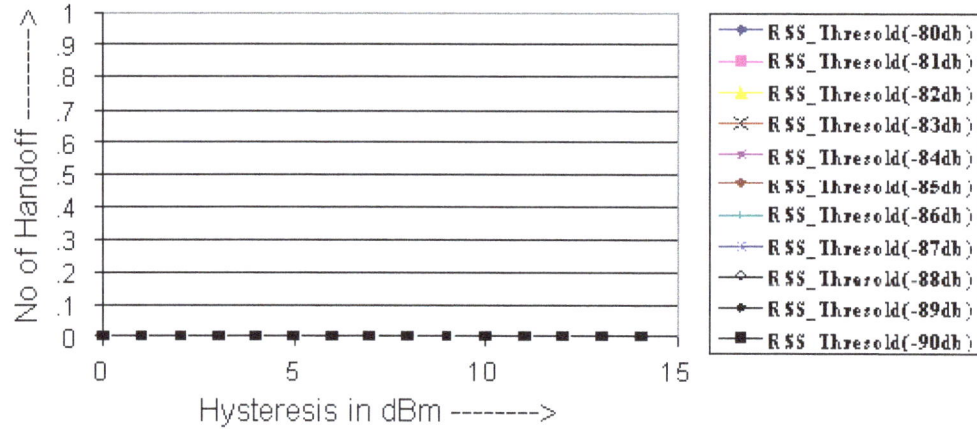

(c)

Fig. 8 No of Handoff vs. different Hysteresis margin for different Threshold value (a) For L/L or M/M or H/H or L/M Traffic Intensities (b) For H/L or H/M Traffic Intensities (c) For L/H or M/H Traffic Intensities

5. CONCLUSIONS

A handoff algorithm using Artificial Neural Network is designed and the performance of the algorithm is studied in this paper. It is observed from the results that the handoff decisions are taken in appropriate positions and the numberof fluctuation are also low. Average number of handoffis also low which minimizes base station and mobile switching centre processor loading. The designed algorithm can be easily embedded and applied to practical mobile cellular networks.

REFERENCES

[1]   N. Benvenuto and F. Santucci, "A least square path loss estimation approach to handoff algorithms", IEEE Transaction on Vehicular Technology, vol. 48, pp 437-447, March 1999.

[2]   Turin, K. L, "Introduction to Spread-Spectrum Antimultipath Techniques and their Application to Urban Digital Radio", Proceedings of the IEEE, vol. 68, no. 3, March 1980, pp. 328 – 353.

[3]   C. Braun, M. Nilson and R. D. Murch, "Measurement of the Interference Rejection Capability of Smart Antennas on Mobile Telephones", IEEE vehicular technology Conference, 1999.

[4]   E. Del. Re, R. Fantacci and G. Giambene, "Handoff and Dynamic Channel Allocation Techniques in Mobile Cellular Networks", IEEE Transactions on Vehicular Technology, Vol 44, No 2, May 1995, pp 229 – 237.

[5]   P. P. Bhattacharya, P. K. Banerjee, "Characterization of Velocity Dependent Mobile Call Handoff", Proc. International Conference on Communication, Devices and Intelligent Systems (CODIS 2004), India, pp. 13-15, 2004.

[6]   B. Venkateswara Rao and Viswanath Sinha, "Study of Channel Assignment Strategies for Handoff and Initial Access in Mobile Communication Networks", IETE Journal of Research, vol 48, No 1, Jan – Feb 2002, pp 69 – 76.

[7]   Traffic Engineering Handbook, Institutes of Transportation Engineering, NOTBO IOA publication, Washington, USA, 1999.

[8]   Sankar Das Mukhopadhyay, "A Comprehensive study of traffic management for the improvement of urban transportation in the districts of West Bengal", Ph.D thesis, Dept. of Business Management, Calcutta University, 2004.

[9]   P. P. Bhattacharya, P. K. Banerjee, "A New Velocity Dependent Variable Hysteresis Margin Based Call Handover Scheme", Indian Journal of radio and Space Physics, Vol. 35, No. 5, pp 368 – 371, October 2006.

[10]   Suleesathira, Raungrong and Kunarak , Sunisa, Non-member,.(2005), "Neural Network Handoff Algorithm in a Joint   Terrestrial-HAPS Cellular System", 164 ECTI transactions on electrical engg., electronics,and communications. VOL.3, NO.2 August 2005

[11]   P. P. Bhattacharya, P. K. Banerjee, "Fuzzy Logic Based Handover Initiation Technique For Mobile Communication In Rayleigh Fading environment", proc. IEEE India annual conference (INDICON 2004), Kharagpur, India, 2004.

[12]   P. P. Bhattacharya, "Application of Artificial Neural Network in Cellular Handoff Management", Proc. International Conference in Computer Intelligence and Multimedia Applications (ICCIMA 07), Sivakasi, India, 2007

12

REALISATION OF AWGN CHANNEL EMULATION MODULES UNDER SISO AND SIMO ENVIRONMENTS FOR 4G LTE SYSTEMS

Dr. R. Shantha Selva Kumari[1] and M. Aarti Meena[2]

[1]Department of Electronics and Communication Engineering, Mepco Schlenk Engineering College, Sivakasi
[2]Department of Electronics and Communication Engineering, Mepco Schlenk Engineering College, Sivakasi

ABSTRACT

The testing of a wireless transmitter and receiver in the real-world channel is tedious. So, a channel emulator using FPGA helps in the testing of transmitter and receiver by providing a test environment that simulates a real-world wireless channel. Since FPGAs are flexible, cheap and reconfigurable, they are used in designing an AWGN channel emulator for 4G LTE for Single Input Single Output (SISO) and Single Input Multiple Output (SIMO) environments. In this paper, three basic modules: transmitter, channel estimation and receiver modules are synthesized. In the transmitter module, the input data is 64 QAM modulated and transmitted into the channel. In the channel estimation module, the transmitter data gets multiplied with the channel coefficients and then added with the noise present in the channel. In th receiver module, the data is detected using MMSE estimation. These are implemented in Virtex-5 device using PlanAhead tool and the Resource and Power Estimations are discussed.

KEYWORDS
AWGN, CFI, Channel coefficients, Channel estimation, MMSE, Pre-computed values, QAM modulation.

1. INTRODUCTION

Channel emulators are of great significance in the testing and verification of wireless communication systems. Design and testing of wireless communication systems using simple tools becomes necessary since the present equipment are not accurate and they cost more. Due to the different characteristics of FPGA like, easy integrated mapping, fast processing, low cost and reconfiguration, the synthesis of channel emulation could be done in an effective way.

Long Term Evolution (LTE) of UMTS (Universal Mobile Telecommunication Systems) is a Fourth Generation wireless broadband technology. The main advantages of LTE are high data rate and low latency with reduced cost, when compared with its predecessor GSM.

In this paper, a channel emulator for AWGN (Additive White Gaussian Noise) is designed for SISO and SIMO environments for LTE systems. They are implemented using PlanAhead tool and Virtex-5 device. The channel is estimated on the basis of BER. Bit Error Rate is defined as the ratio of the error bits to the number of bits transmitted. As the Signal-to-noise ratio increases, BER decreases and vice versa.

2. SYSTEM MODEL

In the transmitter side, it can be observed that the input data is fed into a modulator block. The input data given is four sets of 36 - bits. Those are modulated by means of a 64-QAM modulator. After modulation, the signal is transmitted into the channel, so that it can travel long distances. After modulation, there will be 6 sets of 16 - bit data, sent into the channel.

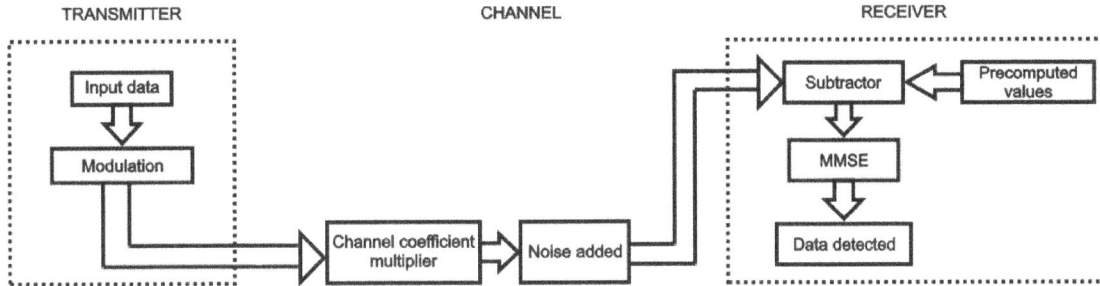

Figure 1. System Model of Channel Emulation process

In the channel, the 6 sets of 16 - bit modulated data are multiplied with the16 sets of 16 - bit channel coefficients which are already present in the channel. Then the resulting data will be added with the noise present in the channel. Channel coefficients are the discrete-time impulse response of the channel. Ideally, one transmitting symbol gives an output only in one symbol time without interfering with other symbols. The 16 sets of 16 - bit data which are already present in the channel are of random in nature. These values are not directly added with the multiplied values. Instead, the 6 sets of 16 - bit noise are multiplied with the variance.

The formula for variance is given by,

$$\text{Variance} = 1 \ \text{sqrt} \ 2 \ * \ 10^{\wedge} \ \text{-SNR} \ 20 \qquad\qquad ....(1)$$

The noise value after getting multiplied with the variance gets added with the multiplied values. This value is received by the receiver.

In the receiver side, the data from the channel is received and it is subtracted from a set of pre-computed values. The pre-computed values are the product of the 6 sets of 16 - bit QAM modulated signal with the 6 sets of 16 - bit channel coefficients. It does not contain the noise which is present in the channel. After subtraction, the resultant value is given as input to the Minimum Mean Square Error estimator (MMSE), to find the minimum value. That minimum value is called as Control Format Indicator (CFI). The data corresponding to the CFI is given as the output by the receiver. The input data is a complex one and hence the output will also be a complex one.

The complex-valued output at the k-th receiving antenna is modelled as,

$$y_k = h_k \circ d(n) + u_k, \quad k = 1, 2, ...K \qquad\qquad ....(2)$$

where, y is the received signal, h is the channel coefficient and u is the noise.
The minimum value will be calculated by,

$$CFI = \min_{m=1,2,3} \sum_{k=1}^{K} \left| y_k - h_k \circ d^m \right|^2 \qquad \ldots\ldots(3)$$

which simplifies to

$$CFI = \arg_{m=1,2,3} z^{(m)} \qquad \ldots\ldots(4)$$

where the soft outputs are given by,

$$z^m = \sum_{k=1}^{K} z_k^m \quad \text{for } m = 1,2,3 \qquad \ldots\ldots(5)$$

where $z_k^m = Re\left(y_k \circ h_k^*, d^m \right)$ for $m = 1,2,3$.

By means of these equations [1], the minimum CFI value is calculated. Then the output retrieved by the receiver is compared with the original data. For every correct value, a counter will count the values. Based on this, BER graph is drawn.

In the case of SISO, there will be a single receiving antenna, whereas in the case of SIMO, there will be two receiving antennae.

3. ARCHITECTURE OF SISO

The architecture of SISO consists of only one transmitting antenna and one receiving antenna as shown in Figure 2.

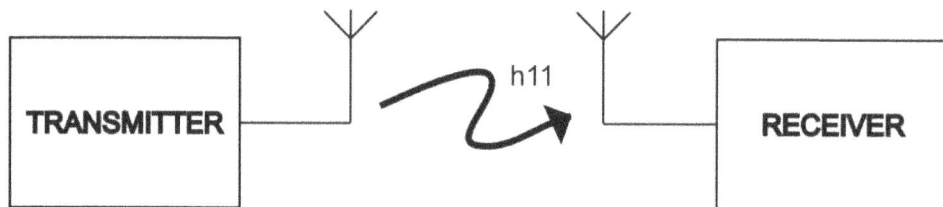

Figure 2. Basic block diagram of SISO

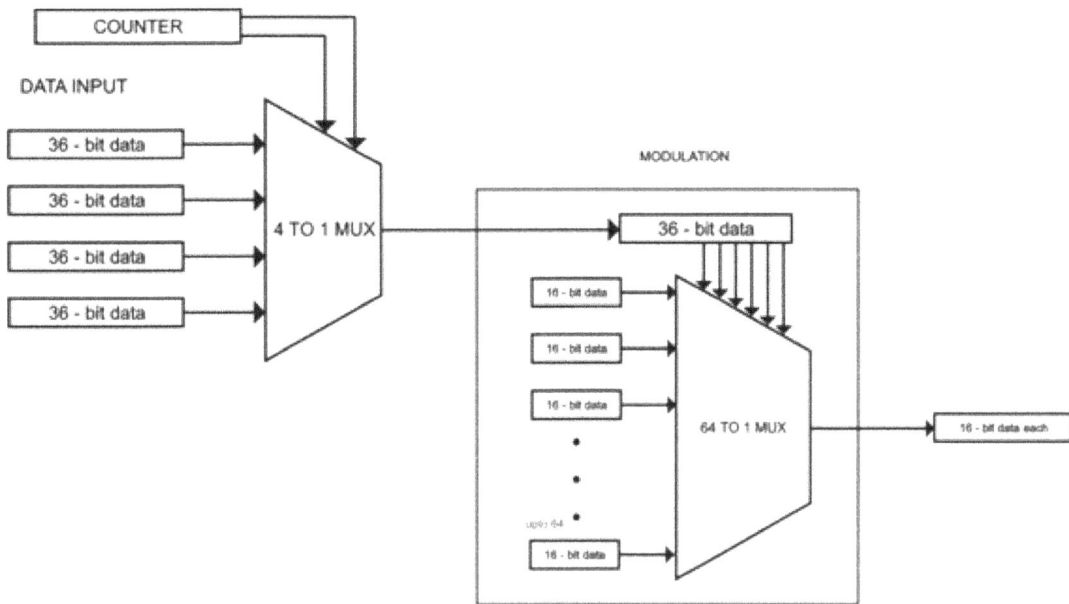

Figure 3. Transmitter side architecture of SISO

From the Figure 3, it is observed that the input given is four sets of 36 - bit data (18 - bit real, 18 - bit imaginary). The output data are given as input to a 4 to 1 MUX. The 2 - bit control input to the MUX are given by a 2 - bit counter. The counter will count for every clock cycle. So for every clock pulse a 36 - bit data will be sent out of the MUX. The next block is a 64-QAM Modulator block. The QAM modulator block consists of four sets of 16 - bit given as input to a 4 to 1 MUX. Each 2 - bit data from the received 36 - bit will act as the control input to the MUX. This is performed by means of a 2 - bit shifter. Based on the values of the 2 - bit data, a 16 - bit data will be given as the output from the modulator. The output will be the QAM modulated signal. Since from a 36 - bit data, modulation takes place for every 2 - bits, the output will be 6 sets of 16 - bit QAM modulated signal. Thus for every clock pulse the output will be 6 sets of 16 - bit QAM modulated signal.

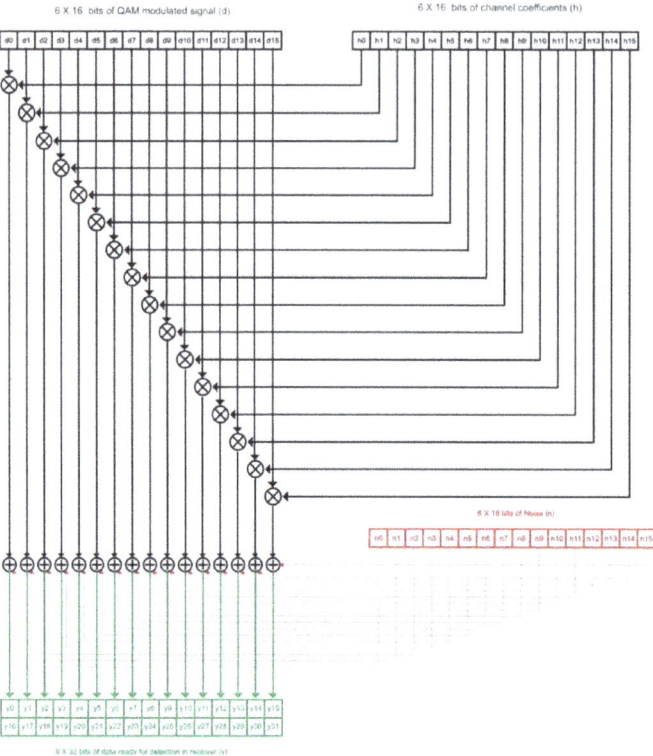

Figure 4. Channel Estimation architecture of SISO

From the Figure 4, it is can be seen that the output from the transmitter side, which is the 6×16 bits QAM modulated signal, is multiplied with the 6×16 bits of channel coefficients [5]. Then the multiplied value is added with the 6×16 bits of noise which is present in the channel. The noise which gets added will be a product of the variance and the noise. The resulting output will be 6×32 bits of data which is to be received by the receiver.

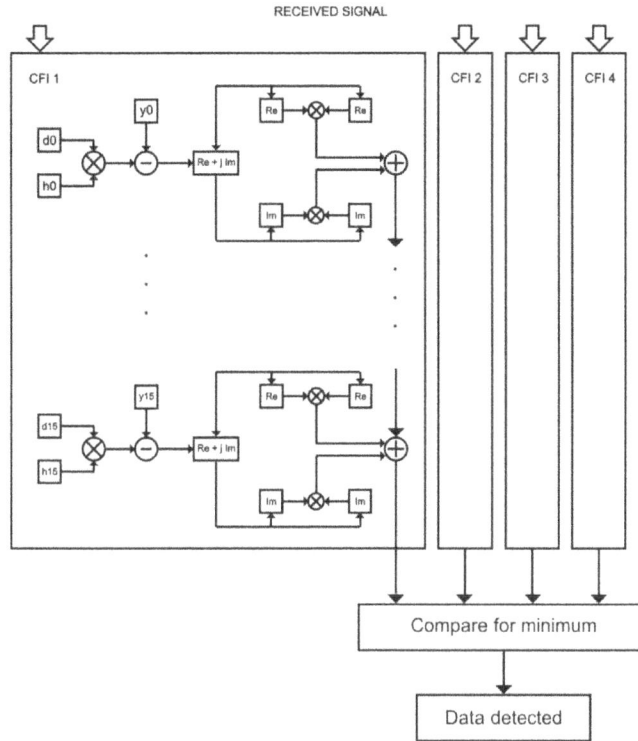

Figure 5. Receiver side architecture of SISO

From the Figure 5, it is observed that in the receiver side, pre-computed data will be present. The pre-computed data is the product of the QAM modulated signal and the channel coefficients value. From the received data, the pre-computed values are subtracted. From the obtained result, real parts and imaginary parts are separately squared and added with each other. This is performed for all the 6 sets of 16 bit data. This is for a single clock pulse. Similarly for all the clock cycles, the values are calculated. Since there are four sets of 36 - bit data, after four clock cycles only the output can be obtained [6]. Hence there will be a delay of four clock cycles. After four clock cycles, there will be four sets of values. These values are compared for minimum, by means of Minimum Mean Square Error (MMSE) estimator algorithm. From this, the minimum value is detected and it is given as the detected output.

4. ARCHITECTURE OF SIMO

The architecture of SIMO consists of only one transmitting antenna and many receiving antennae. In this paper, a 1×2 SIMO is considered (i.e.) one transmitting antenna and two receiving antennae, as shown in Figure 6.

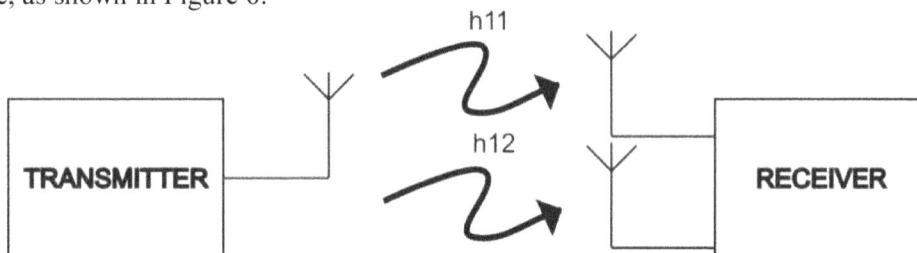

Figure 6. Basic block diagram of SIMO

The transmitter block of SIMO is similar to that of SISO. But the channel estimation and receiver blocks will vary.

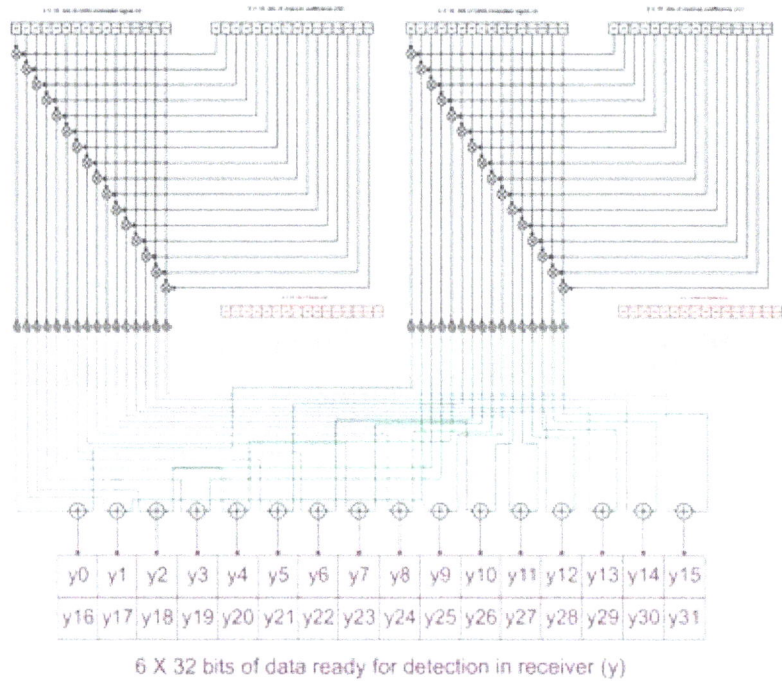

Figure 7. Channel Estimation architecture of SIMO

From the Figure 7, it is observed that the output from the transmitter side, which is the 6×16 bits QAM modulated signal, is multiplied with two sets of 6×16 bits of channel coefficients, which are h0 and h1. After that, there will be two sets of multiplied values. These values are added with two sets 6×16 bits of noise, which are n0 and n1, which is present in the channel. The noise which gets added will be a product of the variance and the noise. Then the two sets of values are added with each other. The resulting output will be 6×16 bits of data which is to be received by the receiver [5].

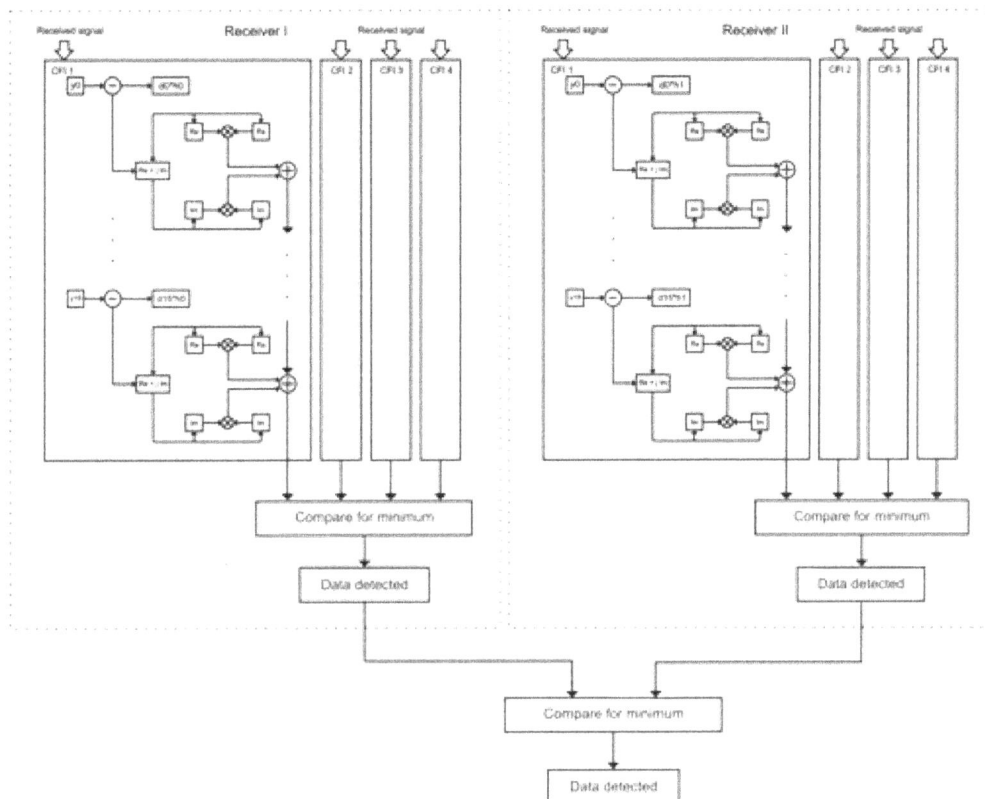

Figure 8. Receiver side architecture of SIMO

In general, in the receiver side, pre-computed data will be present. In the Figure 8, there are two receivers. So, Receiver I will contain pre-computed data which is the product of the QAM modulated signal and the channel coefficients value of h0. In Receiver II, the pre-computed data is the product of the QAM modulated signal and the channel coefficients value of h1. But in both the receivers, the received data will be the same [6].

In each receiver, from the received data, the pre-computed values are subtracted. From the obtained result, real parts and imaginary parts are separately squared and added with each other. This is performed for all the 6 sets of 16 bit data. This is for a single clock pulse. Similarly for all the clock cycles, the values are calculated. Since there are four sets of 36 - bit data, after four clock cycles only the output can be obtained. Hence there will be a delay of four clock cycles.

After four clock cycles, there will be four sets of values, the values of CFI. These values are compared for minimum, by means of Minimum Mean Square Error (MMSE) estimator algorithm. From this, the minimum value is detected and it is given as the detected output. All these processes are executed separately for both the receivers. Finally from the detected output value from both the receivers, the minimum value is calculated. That value is considered as the original data, which is transmitted by the transmitter.

5. RESULTS AND DISCUSSIONS

Simulation is performed in Modelsim and implementation is accomplished by PlanAhead tool in Virtex-5 device.

5.1. Simulation Result

5.1.1. SISO

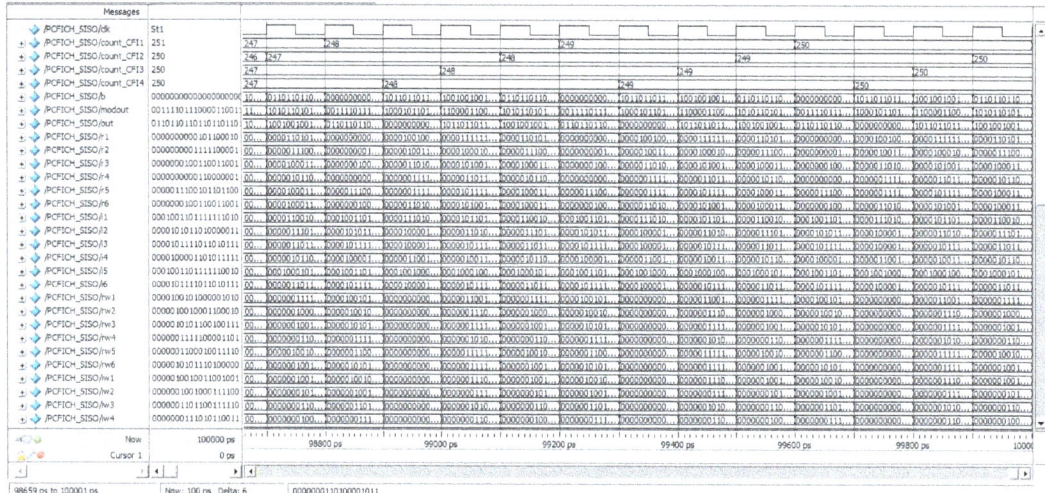

Figure 9. Simulation Result of SISO

From the Figure 9, it is observed that "clk" (clock) is the input given. The outputs count_CFI 1, count_CFI 2, count_CFI 3,count_CFI 4 are the count values which indicate the number of times the original data retrieved at the receiver is equal to the input data given. The Figure 9 indicates the simulation result for a case with variance 0. If the value of variance is 0, then there will be no noise and hence it is the ideal case. So there will be perfect reception of the transmitted signal at the receiver side. However, there will be a delay of four clock pulses and hence each output is counted after four clock pulses. So in the above case, for an input of 1000 clock pulses, the four sets of 36 - bit input data are retrieved 250 times in the Receiver.

5.1.2. SIMO

From the Figure 10, it is observed that "clk" (clock) is the input given. The outputs count1_CFI 1, count1_CFI 2, count1_CFI 3, count1_CFI 4 are the count values which indicate the number of times the original data is retrieved at the receiving end in Receiver I is equal to the input data given in the transmitting end. Similarly, the outputs count2_CFI 1, count2_CFI2, count2_CFI3, count2_CFI4 are the count values which indicate the number of times the original data retrieved at the receiving end in Receiver II is equal to the input data given in the transmitting end.

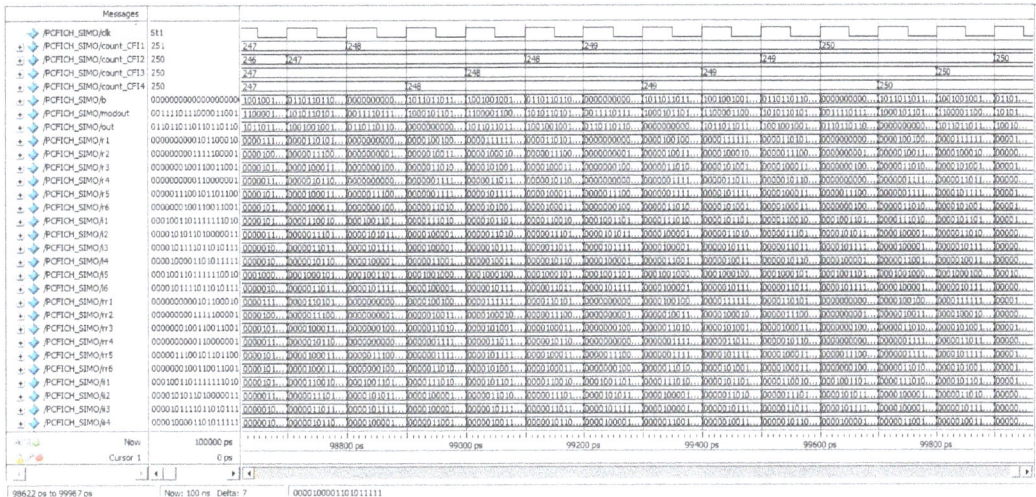

Figure 10. Simulation Result of SIMO

The Figure 10 indicates the simulation result for a case with variance 0. If the value of variance is 0, then there will be no noise and hence it is the ideal case. So there will be perfect reception of the transmitted signal at the receiver side. However, there will be a delay of four clock pulses and hence each output is counted after four clock pulses. So in the above case, for an input of 1000 clock pulses, the four sets of 36 - bit input data are retrieved 250 times in both the Receiver I and Receiver II.

5.2. FPGA Editor

In the FPGA Editor, the programmable elements are called Configurable Logic Blocks (CLB). A CLB contains LUTs (Look Up Table) and (Flip Flops) FFs/Latches. The LUTs and FFs are organized in Slices.

5.2.1. SISO

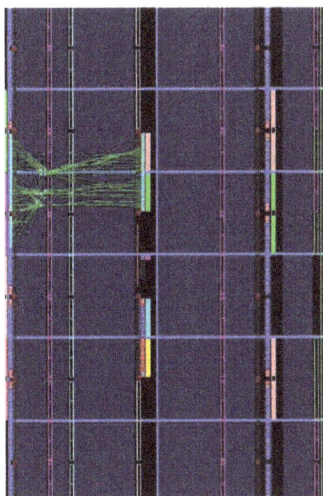

Figure 11. FPGA Editor of Transmitter

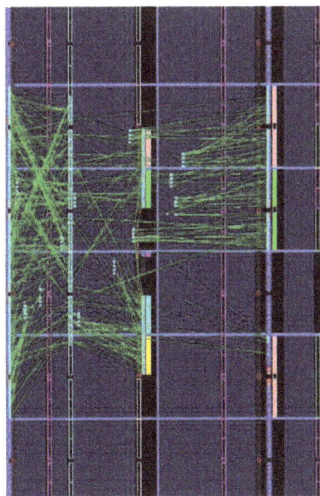

Figure 12. FPGA Editor of Channel

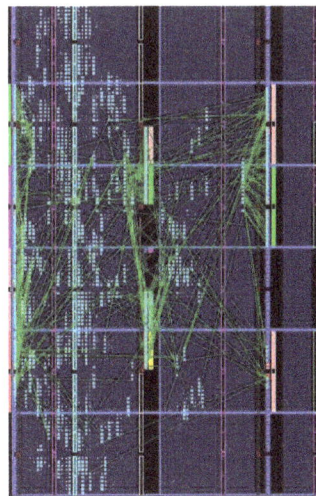

Figure 13. FPGA Editor of Receiver

From the Figure 11, 12 and 13, it can be observed that there are various connecting wires between different LUTs and FFs, which are in green colour. These are the input and output connecting wires.

5.2.2. SIMO

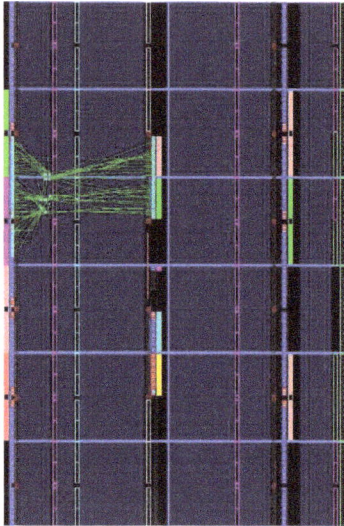

Figure 14. FPGA Editor of Transmitter

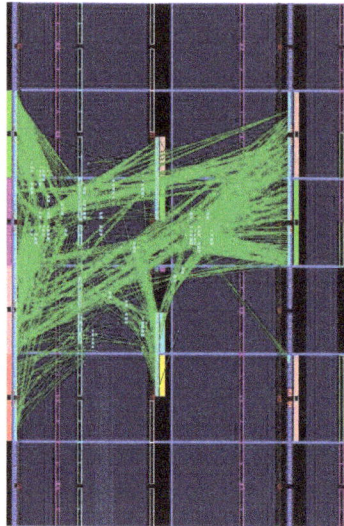

Figure 15. FPGA Editor of Channel

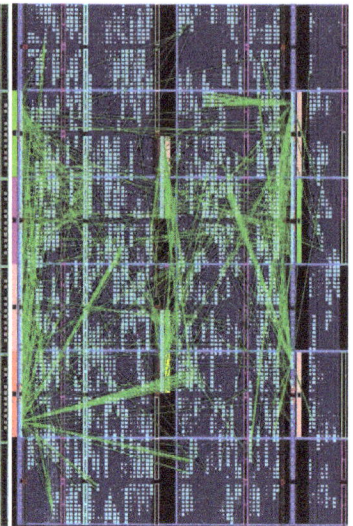

Figure 16. FPGA Editor of Receiver

From the Figure 14, 15 and 16, it can be observed that there are various connecting wires between different LUTs and FFs, which are in green colour. These are the input and output connecting wires.

5.3. RTL Schematic

RTL schematic is a representation of the pre-optimized design in terms of generic symbols, such as adders, multipliers , counters, AND gates, and OR gates.

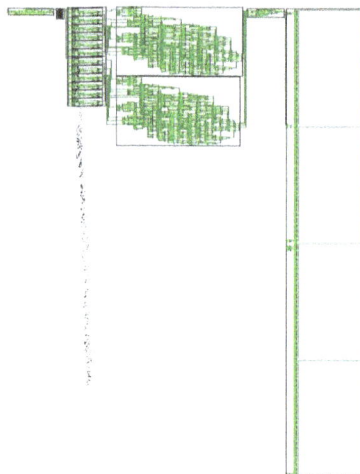

Figure 17. RTL Schematic of SISO

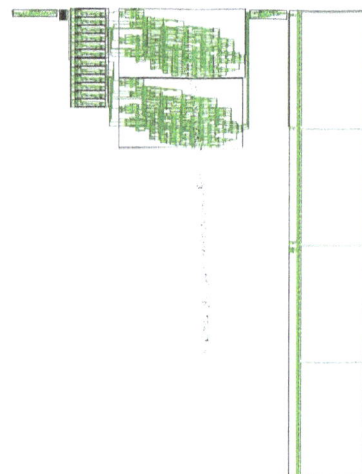

Figure 18. RTL Schematic of SIMO

Figure 17 and 18 shows the RTL Schematic of SISO and SIMO respectively.

5.4. Resource Estimation

Resource estimation indicates the amount of resources that got utilized by the device. In general, it indicates the number of Registers, LUT, Slice, I/O, Buffers utilized.

5.4.1. SISO

Register
Available: 86400
Estimation: 40 (<1% of available) PCFICH_SIMO_Transmitter

LUT
Available: 86400
Estimation: 99 (<1% of available) PCFICH_SIMO_Transmitter

Global Clock Buffer
Available: 32
Estimation: 1 (3% of available) PCFICH_SIMO_Transmitter

IO
Available: 360
Estimation: 97 (27% of available) PCFICH_SIMO_Transmitter

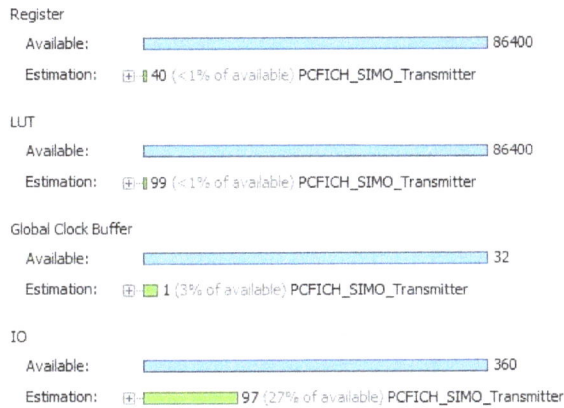

Figure 19. Resource Estimation of Transmitter

From the Figure 19, it can be seen that in the transmitter, the percentage of Registers consumed is 1%, LUT is 1% and IO is 27%.

LUT
Available: 86400
Estimation: 615 (1% of available) PCFICH_SIMO_Channel

DSP48
Available: 48
Estimation: 12 (25% of available) PCFICH_SIMO_Channel

IO
Available: 360
Estimation: 324 (90% of available) PCFICH_SIMO_Channel

Figure 20. Resource Estimation of Channel

From the Figure 20, it is observed that in the channel, the percentage of LUT consumed is 1%, Slice is 25% and IO is 90%.

Register
Available: 86400
Estimation: 100 (<1% of available) PCFICH_SIMO_Receiver

LUT
Available: 86400
Estimation: 13741 (16% of available) PCFICH_SIMO_Receiver

DSP48
Available: 48
Estimation: 48 (100% of available) PCFICH_SIMO_Receiver

Global Clock Buffer
Available: 32
Estimation: 2 (6% of available) PCFICH_SIMO_Receiver

IO
Available: 360
Estimation: 328 (91% of available) PCFICH_SIMO_Receiver

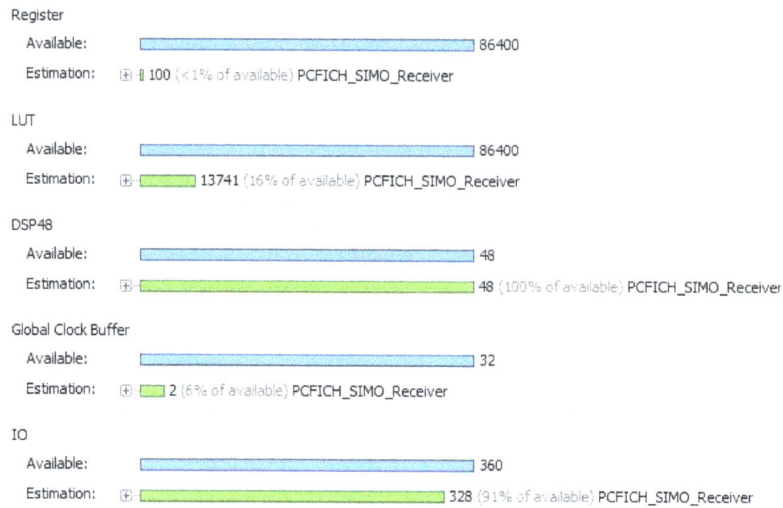

Figure 21. Resource Estimation of Receiver

From the Figure 21, it can be observed that in the receiver side, the percentage of Registers consumed is 1%, LUT is 16% and IO is 91%.

5.4.2. SIMO

Register
Available: 86400
Estimation: 40 (<1% of available) PCFICH_SIMO_Transmitter

LUT
Available: 86400
Estimation: 99 (<1% of available) PCFICH_SIMO_Transmitter

Global Clock Buffer
Available: 32
Estimation: 1 (3% of available) PCFICH_SIMO_Transmitter

IO
Available: 360
Estimation: 97 (27% of available) PCFICH_SIMO_Transmitter

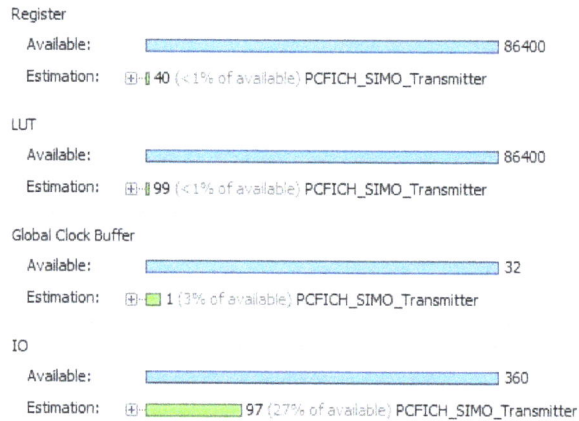

Figure 22. Resource Estimation of Transmitter

From the Figure 22, it can be observed that in the transmitter, the percentage of Registers consumed is 1%, LUT is 1%, Slice is 3% and IO is 27%.

LUT

Available: 86400

Estimation: 615 (1% of available) PCFICH_SIMO_Channel

DSP48

Available: 48

Estimation: 12 (25% of available) PCFICH_SIMO_Channel

IO

Available: 360

Estimation: 324 (90% of available) PCFICH_SIMO_Channel

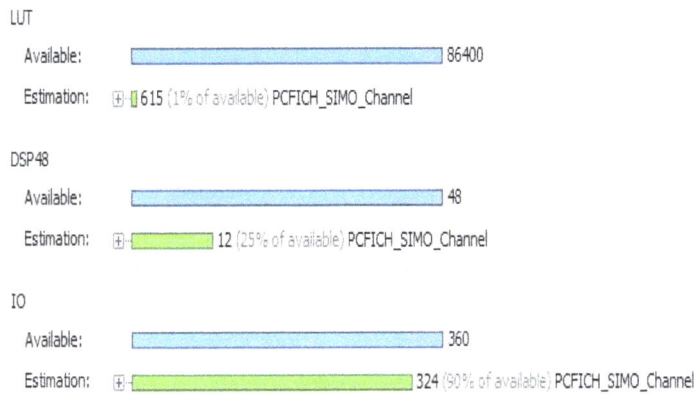

Figure 23. Resource Estimation of Channel

From the Figure 23, it is seen that in the channel, the percentage of Registers consumed is 1%, LUT is 1%, Slice is 25% and IO is 90%.

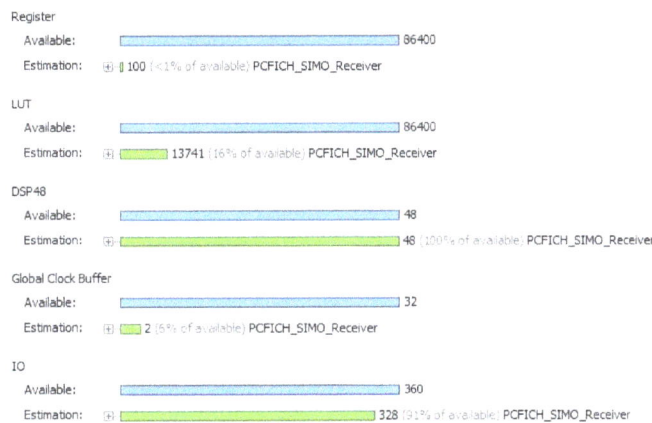

Register

Available: 86400

Estimation: 100 (<1% of available) PCFICH_SIMO_Receiver

LUT

Available: 86400

Estimation: 13741 (16% of available) PCFICH_SIMO_Receiver

DSP48

Available: 48

Estimation: 48 (100% of available) PCFICH_SIMO_Receiver

Global Clock Buffer

Available: 32

Estimation: 2 (6% of available) PCFICH_SIMO_Receiver

IO

Available: 360

Estimation: 328 (91% of available) PCFICH_SIMO_Receiver

Figure 24. Resource Estimation of Receiver

From the Figure 24, it can be observed that in the receiver side, the percentage of Registers consumed is 1%, LUT is 16% and IO is 91%.

5.5. Power Estimation

Power estimation considers the design's resource usage, toggle rates, I/O loading, and many other factors and it combines them with the device models to calculate the estimated power.

There are three types of power estimated in Xilinx:

- I/O Power - It is the power consumed due to external switching.
- Core Dynamic Power - It is the power consumed due to internal switching.
- Device Static Power - It is the power consumed when the device is powered up without programming the used logic. The main contributor of this is the junction temperature.

5.5.1. SISO

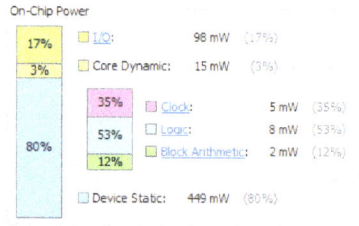

Figure 25. Power Estimation of Transmitter

From the Figure 25, it can be observed that the power consumed by I/O is 98mW, Core Dynamic is 15mW and Device Static is 449mW in the transmitter.

Figure 26. Power Estimation of Channel

From the Figure 26, it is observed that the power consumed by Device Static is 443mW in the channel.

Figure 27. Power Estimation of Receiver

Similarly, from the Figure 27, it can be observed that the power consumed by Device Static is 443mW in the receiver.

5.5.2. SIMO

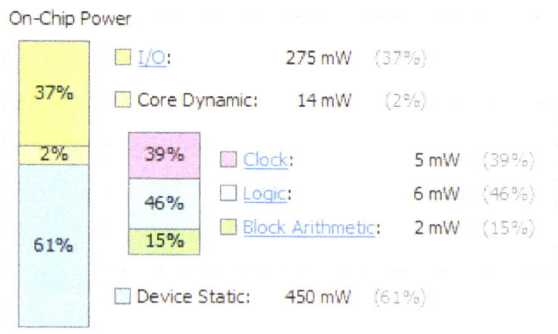

Figure 28. Power Estimation of Transmitter

From the Figure 28, it is observed that the power consumed by I/O is 275mW, Core Dynamic is 14mW and Device Static is 450mW in the transmitter.

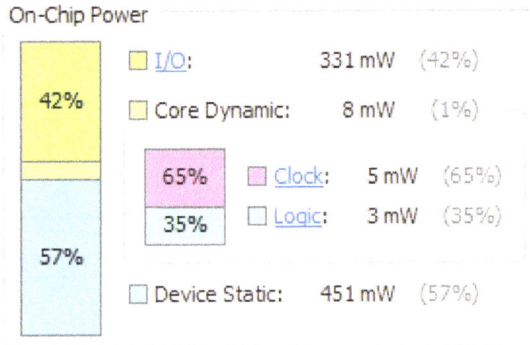

On-Chip Power

I/O: 331 mW (42%)
Core Dynamic: 8 mW (1%)
65% Clock: 5 mW (65%)
35% Logic: 3 mW (35%)
Device Static: 451 mW (57%)

42%
57%

Figure 29. Power Estimation of Channel

From the Figure 29, it is seen that the power consumed by Device Static is 451mW in the channel.

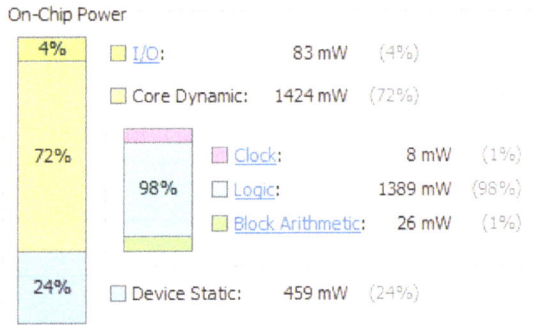

On-Chip Power

4% I/O: 83 mW (4%)
Core Dynamic: 1424 mW (72%)
Clock: 8 mW (1%)
98% Logic: 1389 mW (98%)
Block Arithmetic: 26 mW (1%)
24% Device Static: 459 mW (24%)

72%

Figure 30. Power Estimation of Receiver

Similarly, from the Figure 30, it is observed that the power consumed by Device Static is 459mW in the receiver.

5.6. Timing Summary

The timing summary indicates the overall period, input and output times. It is for all clocks and is limited by the slowest path. In Table 1, various timing parameters of SISO and SIMO are summarised.

Table 1. Timing Summary

Parameters	SISO	SIMO
Speed grade	-2	-2
Minimum period	1.430ns	2.527ns
Maximum frequency	699.301MHz	395.726MHz

5.7. Advanced HDL Synthesis Report

Advanced HDL synthesis report gives the correct technology map and interface to optimize the design. Table 2 and 3 shows the advanced HDL synthesis report of SISO and SIMO.

Table 2. Advanced HDL Synthesis Report of SISO

COMPONENTS	TRANSMITTER	CHANNEL	RECEIVER
ROMs	6	-	2
Multipliers	-	18	48
Adders/Subtractors	1	12	144
Counters	1	-	-
Registers	36	-	-
Latches	-	-	4
Comparators	-	-	6

Table 3. Advanced HDL Synthesis Report of SISO

COMPONENTS	TRANSMITTER	CHANNEL	RECEIVER
ROMs	6	-	2
Multipliers	-	24	96
Adders/Subtractors	1	60	288
Counters	1	-	-
Registers	36	-	-
Latches	-	-	4
Comparators	-	-	6

6. CONCLUSION

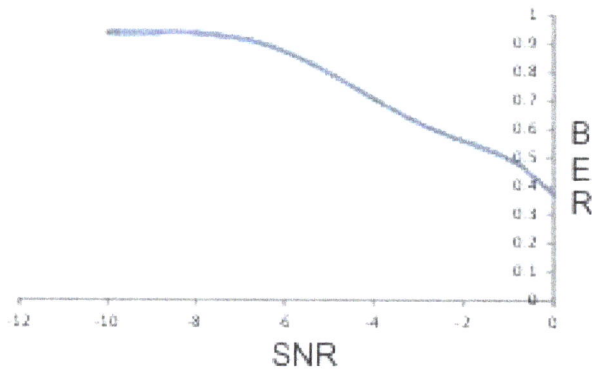

Figure 31. BER Vs. SNR

The output obtained for varying the variance is shown in the Figure 31. From the figure, it is observed that as the Signal-to-Noise Ratio increases, Bit Error Rate decreases. Similarly, channel emulators can be designed for MISO and MIMO environments under different channels like Rayleigh and Rician fading channels and their performance can be compared.

ACKNOWLEDGEMENTS

I would like to thank God, my family and friends for being with me all the time.

REFERENCES

[1]   S. J. Thiruvengadam and Louay M. A. Jalloul, "Performance Analysis of the 3GPP-LTE Physical Control Channels", Hindawi Publishing Corporation, EURASIP Journal onWireless Communications and Networking, Volume 2010, Article ID 914934, 10 pages.\

[2]   Emmanuel Boutillon, Jean-Luc Danger, Adel Ghazel, "Design of high speed AWGN communication channel emulator", IEEE ICECS conference, Kaslik, Lebanon, Dec 2000.

[3]   Jean-Luc Danger, Adel Ghazel, Emmanuel Boutillon, Hédi Laamari, "Efficient implementation of Gaussian noise generator for communication channel emulation", 7th IEEE International Conference on Electronicsm Circuits &Systems (ICECS'2K), Kaslik :Liban (2001).

[4]   Hamid Eslami, Sang V. Tran, Ahmed M. Eltawil, "Design and implementation of a scalable channel emulator for wideband MIMO systems", IEEE Transactions on Vehicular Technology, Vol. 58, No. 9, November 2009.

[5]   Siavash M.Alamouti, "A Simple Transmit Diversity Technique for Wireless Communications", IEEE Journal on Select Areas in Communications, Vol.16, No.8, Oct,1998.

[6]   S. Syed Ameer Abbas, S.J. Thiruvengadam and S. Susithra, "Novel Receiver Architecture for LTE-A Downlink Physical Control Format Indicator Channel with Diversity", Hindawi Publishing Corporation, VLSI Design, Vol. 2014, Article ID 825183, 15 pages.

[7]   S.S.A. Abbas and S.J. Thiruvengadam, "FPGA implementation of 3GPP-LTE-A physical downlink control channel using diversity techniques", International Journal of Wireless and Mobile Computing, Vol. 9, No. 2, p. 84, 2013.

[8]   VLSI Digital Signal Processing Systems: Design and Implementation [book], Keshab K. Parhi.

[9]   Digital Signal Processing with Field Programmable Gate Arrays [book], Uwe Meyer-Bäse.

[10]   Digital Signal Processing: Principles, Algorithms, And Application, 4/E, John G. Proakis.

13

COMPARING THE IMPACT OF MOBILE NODES ARRIVAL PATTERNS IN MANETS USING POISSON AND PARETO MODELS

John Tengviel[1], and K. Diawuo[2]

[1]Department of Computer Science, Sunyani Polytechnic, Sunyani, Ghana
john2001gh@yahoo.com
[2]Department of Computer Engineering, KNUST, Kumasi, Ghana
kdiawuo@yahoo.com

ABSTRACT

Mobile Ad hoc Networks (MANETs) are dynamic networks populated by mobile stations, or mobile nodes (MNs). Mobility model is a hot topic in many areas, for example, protocol evaluation, network performance analysis and so on.How to simulate MNs mobility is the problem we should consider if we want to build an accurate mobility model. When new nodes can join and other nodes can leave the network and therefore the topology is dynamic.Specifically, MANETs consist of a collection of nodes randomly placed in a line (not necessarily straight). MANETs do appear in many real-world network applications such as a vehicular MANETs built along a highway in a city environment or people in a particular location. MNs in MANETs are usually laptops, PDAs or mobile phones.

This paper presents comparative results that have been carried out via Matlab software simulation. The study investigates the impact of mobility predictive models on mobile nodes' parameters such as, the arrival rate and the size of mobile nodes in a given area using Pareto and Poisson distributions. The results have indicated that mobile nodes' arrival rates may have influence on MNs population (as a larger number) in a location. The Pareto distribution is more reflective of the modeling mobility for MANETs than the Poisson distribution.

KEYWORDS

Mobility Models, MANETs, Mobile Nodes Distribution, Arrival Patterns, Pareto Distribution, Poisson Distribution, Matlab Simulation.

1.INTRODUCTION

Mobile Ad-hoc NETworks (*MANETs*) is a collection of wireless mobile nodes configured to communicate amongst each other without the aid of an existing infrastructure. MANETS are *Multi-Hop* wireless networks since one node may not be indirect communication range of other node. In such cases the data from the original sender has to travel a number of hops (hop is one communication link) in order to reach the destination. The intermediate nodes act as routers and forward the data packets till the destination is reached [1].

Recently, with the deployment of all kinds of wireless devices, wireless communication is becoming more important. In this research area, Ad-Hoc network is a hot topic which has attracted much of research attentions. A wireless ad hoc network is a decentralized wireless network. The network is ad hoc because it does not rely on a preexisting infrastructure, such as routers in wired networks or access points in managed (infrastructure) wireless networks. Instead, each node participates in routing by forwarding data for other nodes, and so the determination of which nodes forward data is made dynamically based on the network connectivity [2]. There are different kinds of routing protocol defined by how messages are sent from the source node to the destination node.

Based on this, it's reasonable to consider node mobility as an essential topic of ad-hoc network. With an accurate mobility model which represents nodes movement, designers can evaluate performance of protocols, predict user distribution, plan network resources allocation and so on. It can also be used in healthcare or traffic control area rescue mission, and so on.

Ad hoc networks are viewed to be suitable for all situations in which a temporary communication is desired. The technology was initially developed keeping in mind the military applications [3] such as battle field in an unknown territory where an infrastructure network is almost impossible to have or maintain. In such situations, the ad hoc networks having self-organizing [4] capability can be effectively used where other technologies either fail or cannot be effectively deployed. The entire network is mobile, and the individual terminals are allowed to move freely. Since, the nodes are mobile; the network topology is thus dynamic. This leads to frequent and unpredictable connectivity changes. In this dynamic topology, some pairs of terminals may not be able to communicate directly with each other and have to rely on some other terminals so that the messages are been delivered to their destinations. Such networks are often referred to as multi-hops or store-and-forward networks [5].

This paper presents a study on mobile nodes arrival patterns in MANETs using Poisson and Pareto models. Though not very realistic from a practical point of view, a model based on the exponential distribution can be of great importance to provide an insight into the mobile nodes arrival pattern. The section 2 illustrates a brief review on MANETs studies. The section 3 introduces the Poisson and Pareto distribution models. The simulation procedures and considered parameters are presented in section 4. The obtained results are objects in section 5 and the section 6 closes the paper to further research works.

2. RELATED WORKS

Currently there are two types [6, 7]of mobility models used in simulation of networks. These are traces and synthetic models. Traces are those mobility patterns that are observed in real-life systems. Traces provide accurate information, especially when they involve a large number of mobile nodes (MNs) and appropriate long observation period. On the other hand, synthetic models attempt to realistically represent the behaviour of MNs without the use of traces. They are divided into two categories, entity mobility models and group mobility models [1, 8, 9]. The entity mobility models randomise the movements of each individual node and represent MNs whose movements are independent of each other. However, the group mobility models are a set of groups' nodes that stay close to each other and then randomise the movements of the group and represent MNs whose movements are dependent on each other. The node positions may also vary randomly around the group reference point. In [10], the mobility study in ad hoc has been approximated to pedestrian in the street, willing to exchange content (multimedia files, mp3, etc.) with their handset whilst walking at a relative low speed. Some researchers have proposed basic mobility models such as Random Walk, Random Waypoint, [3, 4], etc. for performance comparison of various routing protocols. The concern with these basic designed models is that they represent a specific scenarios not often found in real lives. Hence their use in ad hoc network studies is very limited. Random Walk or Random Waypoint model though simple and elegant, produce random source of entry into a location with scattered pattern around the simulation area, sudden stops and sharp turns. In real-life, this may not really be the case.

3. MODELS OF STUDY

3.1. POISSON ARRIVAL DISTRIBUTION (NUMBER OF NODES)

When arrivals occur at random, the information of interest is the probability of n arrivals in a given time period, where $n = 0, 1, 2, \ldots\ldots n-1$

Let λ be a constant representing the average rate of arrival of nodes and consider a small time interval Δt, with $\Delta t \to 0$. The assumptions for this process are as follows:

- The probability of one arrival in an interval of Δt seconds, say **(t, t+Δt)** is $\lambda\Delta t$, independent of arrivals in any time interval not overlapping **(t, t+Δt).**
- The probability of no arrivals in Δt seconds is **1-$\lambda\Delta$t**, under such conditions, it can be shown that the probability of exactly **n** nodes arriving during an interval of length of **t** is given by the Poisson distribution law [11] in equation 1:

$$\mathbf{P(n)} = \frac{(\lambda t)^n e^{-\lambda t}}{n!}, \quad \text{where } n \geq 0, \ t > 0. \quad (1)$$

The assumption of Poisson MN arrivals also implies a distribution of the time intervals between the arrivals of successive MN in a location.

3.2. Pareto Distribution

The Pareto distributions [12-14] are characterized by two parameters: α and β. Parameter α is called shape parameter that determines heavy-tailed characteristics and $\beta = 1$ is called cutoff or the location parameter that determines the average of inter-arrival time.

The node arrival times of the Pareto distribution are independent and identically distributed, which means that each arrival time has the same probability distribution as the other arrival times and all are mutually independent. The two main parameters of the Pareto process are the shape α and the scale parameter (x).

For one parameter Pareto (α shape only), the distribution function can be written as equation 2:

$$F(X) = 1 - \left(\frac{1}{1+X}\right)^{\alpha}, X \geq 0 \quad (2)$$

The pdf is given as in equation 3:

$$f(X) = \frac{\alpha}{(1+X)^{\alpha+1}}$$ (3)

and for the two – parameter Pareto distribution function defined over the real numbers can be written as in (4):

$$\begin{cases} F(X) = 1 - \left(\frac{1}{\alpha+X}\right)^{\beta} \\ \qquad , \\ X \geq 0; \quad \alpha, \beta > 0 \end{cases}$$ (4)

Its pdf is given as in equation 5:

$$f(X) = \frac{\alpha}{\beta} * \left(\frac{\beta}{X}\right)^{\alpha}$$ (5)

4. METHODOLOGY

4.1. Varying of α in Pareto Arrival Distribution

We assume the arrival distribution on the MNs population by using Pareto distributions.

Table 1: Varying α parameter values

Scenario	1	2	3	4	5
α (B)	0.3	0.4	0.5	0.8	0.9

For the simulations purposes, the varying α values are been considered. Heavy-tail is been modeled by a Pareto distribution and the main principle can be attributed to the principle of number of nodes. We have performed the simulations for a wide range of parameter values as in Table 1 for both one-parameter and two-parameter Pareto models.

4.2. VARYING OF ARRIVAL RATES FOR NODE DISTRIBUTION

The arrival pattern of mobile nodes has an impact on the performance of the network. In this scope, we have decided to analysis the effect of arrival distribution on the MNs population in a given area by using Poisson distribution as in equation 1.In most real-world MANETs, the node population in an area of interest varies with time. In this simulation, it is therefore necessary to investigate the impact of arrivals of MNs on the MANETs mobility.

The simulation area does not change as the arrival rate changes. The different values of arrival rates being considered in this study are shown in Table 2.

Table 2: Varying Arrival Rates

Scenario	1	2	3	4	5
Arrival rates	0.3	0.4	0.5	0.8	0.9

During the simulation, nodes were allowed to enter the location from a common source (0 degrees) but not from different sources. The number of MNs that entered the location was assumed to be Poisson distributed with varying arrival rates.

5. RESULTS AND DISCUSSION

5.1. Comparative Study using Pareto Arrival Pattern

In this section, the effect of arrival rates on MNs distribution and population in a defined location is analyzed as shown in Figure 1. It was observed that the various arrival rates increased the number of MNs also increased but to a certain limit. It is therefore the indication that every location has a limit or capacity of MNs it can contain.

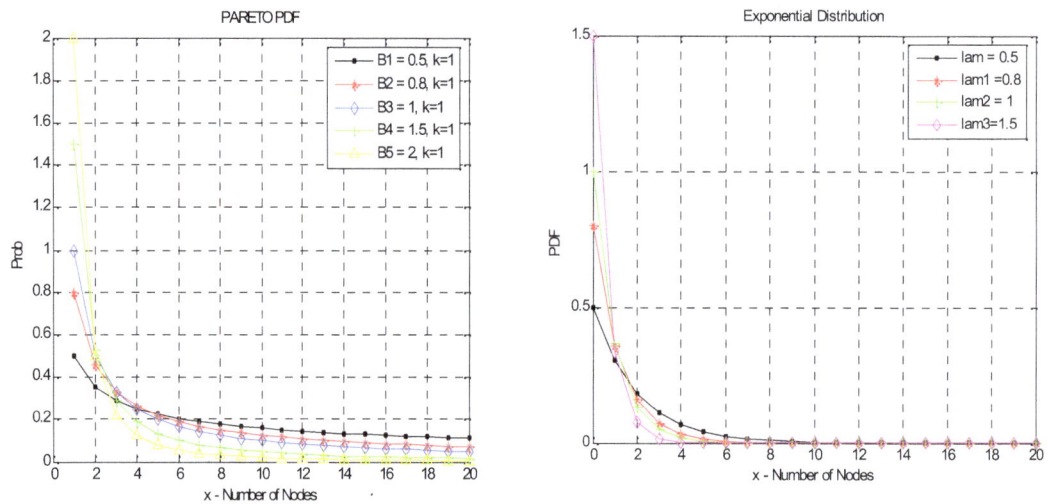

Figure 1: Single Parameter for Varying Values for B, and Exponential for Twenty Nodes

Figure 1may indicate that the exponential distribution was higher than the single parameter in the initial stages but as time progresses the exponential decreases fast to zero. The single parameter Pareto overtakes the exponential as the number of nodes increases and indication that the single parameter performs better than exponential distribution.

The Pareto distribution may show tail that decays much more slowly than the exponential distribution. The alpha is the shape parameter which determines the characteristics "decay" of the distribution (tail index) and A is the location parameter which defines the minimum value of x (number of nodes).

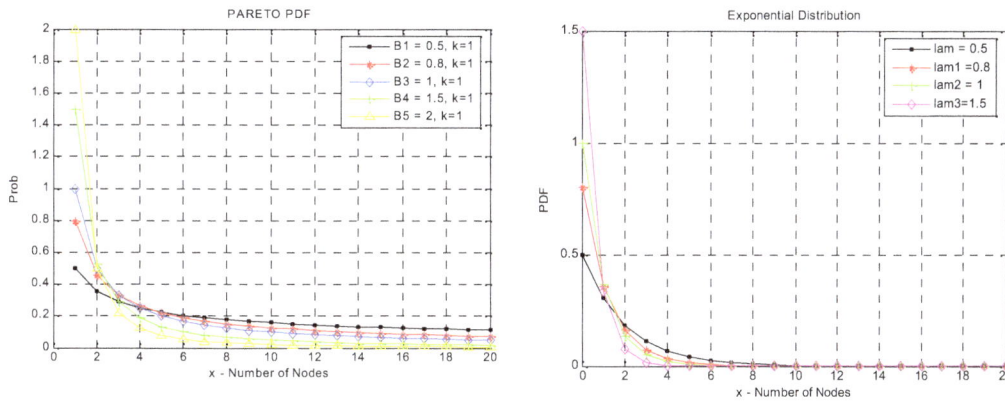

Figure 2: Two Parameter Pareto for Varying B Values and Exponential

In Figure2 the comparison between the two-parameter Pareto and exponential distributions is illustrated. It is obvious that the two-parameter Pareto outweighs the exponential distribution as the number of MNs increases. The exponential distributions decays very fast and finally get to the a-axis unlike the two-parameter Pareto distribution where some of the arrival rates distribution has not decay to zero.

However the two-parameter Pareto performed well than the one-parameter Pareto, since some of the arrival of the two-parameter did not decay to zero. The long-tailed nature of the two-parameter Pareto helped to clear out any congestion in a location when the arrival rate was small and the reverse was also true.

5.2 Effect of Varying Arrival Rates

In Figure 3, the effect of varying nodes' arrival rate is computed using Poisson model. Nodes may arrive at a location either in some regular pattern or in a totally random fashion. The arrival rates have shown to impact on the number of nodes in a particular location, although every location has a limited capacity. A high number of nodes typically translate into a higher average number of neighbours per node, which influences the route availability.

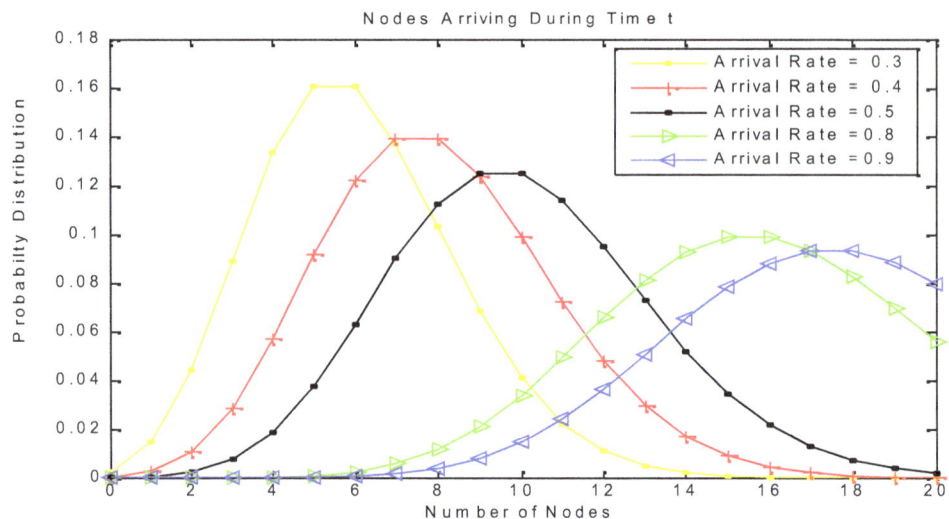

Figure 3: For Twenty Number of Nodes for varying Arrival rates

In reality, the total connection time of a node over a specific interval depends on the nodes encounter rate and the time in each encounter, both of which depend on the relative mobility of nodes.

Although a high node arrivals results in more node encounters, the network would eventually become congested. The impact of this relationship is that nodes can and will be tightly packed (i.e. High density) if their arrival rates is high (congestion), but if the arrivals is lower, the nodes must be farther apart (low density). For instance it is clear that there is some congestion for arrivals of MNs, since they have to follow some holding paths.

As the value of arrival rate increases, the shape of the distribution changes dramatically to a more symmetrical ("normal") form and the probability of a larger number of arrivals increases with increasing number of MNs. An interesting observation is that as the arrival rate increases, the properties of the Poisson distribution approach those of the normal distribution as in Figure3.
The first arrival processes of nodes give higher contact probabilities at higher arriving rates. This is due to the nodes' contiguity one to another making mobility difficult. In practice, one may record the actual number of arrivals over a period and then compare the frequency of distribution of the observed number of arrival to the Poisson distribution to investigate its approximation of the arrival distribution.

6. CONCLUSION

The arrival patterns have shown some impact on the network population, as the arrival rate increases the MNs population also increases to a peak and then decays rapidly to the x-axis. It was realized that the Poisson distribution is not good for the arrival distribution; therefore the Pareto distribution was considered. It has come out clear that the Pareto distribution is good for the arrival distribution, especially the two-parameter Pareto distribution which performed better than the single Pareto and exponential distributions even though at the earlier stages the exponential performed than the single Pareto distribution with a faster decay.

It may subsequently be admitted that mobility in MANETs is a difficult work and actually. It is an interesting research area that has been growing in recent years. Its difficulty is mainly generated because of the continuous changes in the network topology with time. The topological changes have impact on mobility techniques developed for infrastructure-based networks thus may not be directly applied to mobile adhoc networks. We have investigated through simulation mobility prediction of MNs using the queueing model.

REFERENCES

[1]   J. Boleng, T. Camp, and V. Tolety. "A Suvey of Mobility Models for Ad hoc Network Research", In Wireless Communication and Mobile Computing (WCMC), Vol. 2, No. 5, pages 483 – 502, 2002.

[2]   Subir Kumar Sarkar, T. G. Basavaraju, C. Puttamadappa, "Mobility Models for Mobile Ad Hoc Networks", 2007, Page 267 – 277, Auerbach Publications – www.auerbach-publications.comVolume 2, No.5, 2002.

[3]   C. Rajabhushanam and A. Kathirvel, "Survey of Wireless MANET Application in Battlefield Operations", (IJACSA) International Journal of Advanced Computer Science and Applications, Vol. 2, No.1, January 2011.

[4]   Buttyan L., and Hubaux J. P., "Stimulating cooperation in self-organizing mobile ad hoc networks. Mobile Networks andApplications: Special Issue on Mobile Ad Hoc Networks, 8(5), 2003.

[5]   C.P.Agrawal, O.P.Vyas and M.K Tiwari, "Evaluation of Varying Mobility Models & Network Loads on DSDV Protocol of MANETs", International Journal on Computer Science and Engineering Vol.1 (2), 2009, pp. 40 - 46.

[6]   P. N. Pathirana, A. V. Savkin& S. K. Jha. "Mobility modeling and trajectory prediction for cellular networks with mobile base stations". MobiHoc 2003: 213 -221.

[7]   MohdIzuanMohdSaad and Zuriati Ahmad Zukarnain, "Performance Analysis of Random-Based Mobility Models in MANET Routing Protocol, EuroJournals Publishing, Inc. 2009, ISSN 1450-216X Vol.32 No.4 (2009), pp.444-454 http://www.eurojournals.com/ejsr.htm

[8]   Zainab R. Zaidi, Brian L. Mark: "A Distributed Mobility Tracking Scheme for Ad-Hoc Networks Based on an Autoregressive Model". The 6th International Workshop of Distributed Computing, Kolkata, India (2004) 447(458)

[9]   Abdullah, SohailJabbar, ShafAlam and Abid Ali Minhas, "Location Prediction for Improvement of Communication Protocols in Wireless Communications: Considerations and Future Directions", Proceedings of the World Congress on Engineering and Computer Science 2011 Vol. II WCECS 2011, October 19-21, 2011, San Francisco, USA

[10]   Gunnar Karisson et al., "A Mobility Model for Pedestrian Content Distribution", SIMUTools '09 workshops, March 2-6, 2009, Rome Italy.

[11]   John Tengviel, K. A. Dotche and K. Diawuo, "The Impact of Mobile Nodes Arrival Patterns In Manets Using Poisson Models", International Journal of Managing Information Technology (IJMIT), Vol. 4, No. 3, August 2012, pp. 55 – 71.

[12]   Martin J. Fischer and Carl M. Harris, "A Method for Analysing Congestion In Pareto and Related Queues", pp. 15 – 18.

[13]   K. Krishnamoorthy, "Handbook of Statistical Distributions with Applications", University of Louisiana at Lafayette, U.S.A. pp. 257 – 261.

[14]   Kyunghan Lee, Seongik Hong, Seong Joon Kim, Injong Rhee and Song Chong, "SLAW: Self-Similar Least-Action Human Walk", published in the Proceedings of the IEEE Conference on Computer Communications (INFOCOM), Rio de Janeiro, Brazil, April 19–25, 2009.. PP. 855 – 863.

14

Design and Performance Evaluation of OFDM-Based Wireless Services Employing Radio over Optical Wireless Link

Jiang Liu, Wasinee Noonpakdee, Shigeru Shimamoto

Global Information and Telecommunication Institute, Waseda University, 1-3-10, Nishi-Waseda, Shinjuku-Ku, Tokyo, 169-0051 Japan

liujiang@aoni.waseda.jp, wasinee@toki.waseda.jp, shima@waseda.jp

ABSTRACT

This paper studies an Orthogonal Frequency Division Multiplexing (OFDM) based wireless services employing Radio over Optical Wireless (RoOW) to get a high performance transmission while eliminating the drawback of possible radio wave interference to electro-medical apparatus in indoor communication system. An optical Intensity-modulated Direct-detection (IM/DD) system using RF subcarrier modulation is considered and OFDM RF signals are assumed as the subcarrier. The transmission performance is evaluated by simulation considering the influence of peak clip of OFDM signal and different radio fading scenarios. Simulation result shows that the modulation index m, optical transmit power, radio environments are important factors which affect the communication performance significantly. It also shows when the received optical power is larger than -20dBm, which is a very practical power level, the value of BER becomes almost independent of the optical power and the radio environment becomes more important. That means the indoor optical wireless communication (OWC) has high transmission performance and the proposed relay system is a very available system as long as the radio environment is eligible. We believe that the proposed system is a practical system and can provide a low-cost, high-quality wireless services for RF sensitive areas.

Keywords

Optical Wireless Communication, Relay Station, Intensity-modulated Direct-detection (IM/DD), RF Subcarrier, Diversity

1. INTRODUCTION

Mobile terminals including wireless voice telephone, mobile internet access, video calls and mobile TV have become more and more necessary in our daily life. Right from a school going child, a house wife to a servant, the mobile terminals has its major impact on their lives [1]. Users want to use them without any spatial constraint because they have become indispensable tools. However, the electromagnetic interference with electro-medical apparatus generated by using radio frequency (RF) communication becomes a health issue with the increasing number of mobile users. In medical facilities, such as hospitals, clinics, and doctor's offices, sensitive equipments may malfunction if exposed to high RF powers, possibly endangering patients' lives [2]-[4]. Therefore, in medical institutions, mobile phones are required to be turned off except at some special designated places. Similarly, all railroad companies demand that mobile phones should be turned off nearby priority seats. However there are so many mobile phone users that are difficult to be convinced to turn off their mobile phones.

Optical wireless communication (OWC) system is safe to electro-medical apparatus and can keep the convenience of current wireless communications using radio waves. Moreover, OWC has features like unlimited license-free bandwidth, cheaper transceivers, high security etc [5]-[10], which is suitable to RF sensitive service areas. As a feasible application of OWC, we propose a relay scheme for an indoor optical wireless communication system tends to provide non-RF communication for the last few meters to/from mobile terminals operating in an RF-sensitive area. It also can provides applications for the dead zones where wireless signal from base station cannot reach or is very weak such as tunnel, areas behind buildings or mountainous places etc.

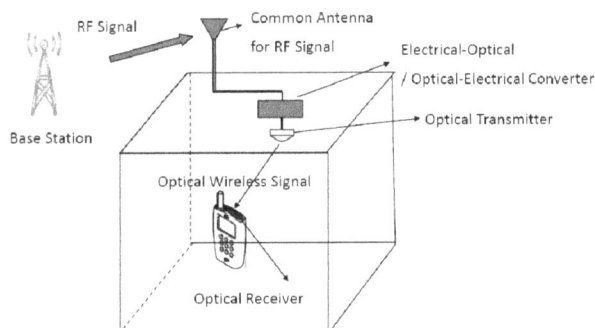

Fig. 1 Relay System Model

Orthogonal frequency division multiplexing (OFDM) is one of the most popular techniques for high data rate wireless communication and is great interested by the researchers in the universities and research laboratories all over the world. It is an effective high speed data transmission scheme without using very expensive equalizers and it has been proposed as the air interface for broadband wireless applications [11]. OFDM has been accepted in the IEEE802.11a local area network (LAN), IEEE802.16a local area network /Metropolitan area network (LAN/MAN), High Performance Local Area Network type 2 (HIPERLAN/2), standards, and Mobile Multimedia Access Communication (MMAC) Systems. OFDM is also being considered in IEEE802.20a, a standard in the making for maintaining high-bandwidth connections to users moving at speeds up to 60 mph[12], [13]. OFDM is considered a potential candidate for the next-generation mobile wireless systems [14], [15].

This paper presents an OFDM-Based Wireless Services for indoor RF sensitive area. One of the main disadvantages of OFDM is its high peak-to-average power ratio (PAPR) [11] and high amplitude peaks may be clipped. This system is quantitatively evaluated considering the peak clip. Moreover, the BER performance is studied in different fading scenarios.

The remainder of this paper is organized as follows. System model is introduced in section 1. The related work is given in section 2 and the system model is discussed in section 3. In Section 4, we present the mathematical modeling for the transmission of OFDM signals over RoOW link. The analytical results are shown and discussed in Section 5. Finally, Section 6 concludes the paper.

2. RELATED WORK

There are two methods which can be considered to relay RF signals to mobile terminal inside the service area. One is demodulating the information data to baseband signal and use optical digital modulation schemes such as on-off-keying (OOK) modulation. However, each mobile user in a service area may use different RF broadband due to have a contract with different provider. It is difficult to perform the regeneration process at the relay station. Moreover, multiple access schemes have to be considered for multiple user case. Another relay method is using non-regeneration system. If non-regeneration system is used, only a minimal set of front-end hardware components is needed in the receiver/transmitter functions at the relay station. In this paper, a non-regeneration system is adopted. We propose an optical intensity-modulated direct-detection (IM/DD) [16] relay scheme for an indoor optical wireless communication system, referred to as Radio over Optical Wireless (RoOW).

Radio transmission over optical fiber links, referred to as radio-over-fiber (RoF), has been viewed as a way of simplifying the architecture of remote antenna base-stations and realizing high-performance networks [17]. Similar to RoF, Transmission of RF using OWC link provides as a way of simplifying the architecture of relay station. It combines the advantages of high transmission capacity enabled by optical device technologies and easy deployment of wireless links. By using the proposed RoOW scheme, at the relay station, the received RF signals are directly modulated with IM, up-converted into optical band, and propagated as optical signals. The RF signal is defined to be RF subcarrier in this paper. The details of the proposed system model will be introduced in next section.

3. SYSTEM MODEL

In this paper, we propose a relay system using OWC as shown in Figure 1. This model proposes to set up a relay station for mobile phone coverage inside medical institutions, trains, apartments, etc. It is proposed to keep the present wireless access from base station, and replace the radio signal link between the relay station and the mobile terminal with optical wireless signals. The conventional RF wireless communication is used between the relay station and the base station. For the relay station, the RF transceiver is set up outside the service area, and the optical transceiver is set up on the ceiling or the wall inside the service area.

We propose to use OWC both for downlink and uplink. The wavelength of the optical wireless signal used in the downlink is different from the wavelength of optical wireless signal used in the uplink. The downlink from the base station to the mobile terminal is established in two steps (refer to Figure 1). Firstly the relay station receives RF signals; secondly the received RF signals are converted into optical signals and sent to the mobile terminal using light emission elements. In the uplink, mobile terminal convert RF signal to optical signals and send it to optical receivers inside the room. The RF signals are extracted from the received optical wireless signals in the relay station, and then sent to the base station as uplink signals. The RF signals sent from the relay station can be sent at a higher power level compared with the RF signals sent directly from the mobile terminal to achieve higher transmission quality. Moreover, in the downlink, better than RF relay method, if mobile terminal can also receive some RF signals from the base station, there are no interference between the RF signals and the optical wireless signals due to different transmission band.

Since RF signals are not converted to base band signals in the proposed non-regenerative relay station, the circuits added to the relay station can be reduced. Therefore, we believe that the proposed system using such a combination can save electromagnetic interference with electro-medical equipment in the coverage area without changing the present communication infrastructure.

4. Mathematical modelling for the transmission of OFDM signals over RoOW link

We consider an optical IM/DD system using RF subcarrier modulation and OFDM signal is assumed as the RF subcarrier signal. The configuration of OFDM signal transmission over optical wireless link is shown as Figure 2. An OFDM carrier signal is the sum of a number of orthogonal sub-carriers, with baseband data on each sub-carrier being independently modulated commonly using some type of quadrature amplitude modulation (QAM) or phase-shift keying (PSK). In this paper, QPAK modulation is assumed. A serial data stream of binary digits is first multiplexed into parallel streams, and each one mapped to a symbol stream using QPSK modulation constellation. IFFT and FFT are implemented at the transmitter and the receiver respectively.

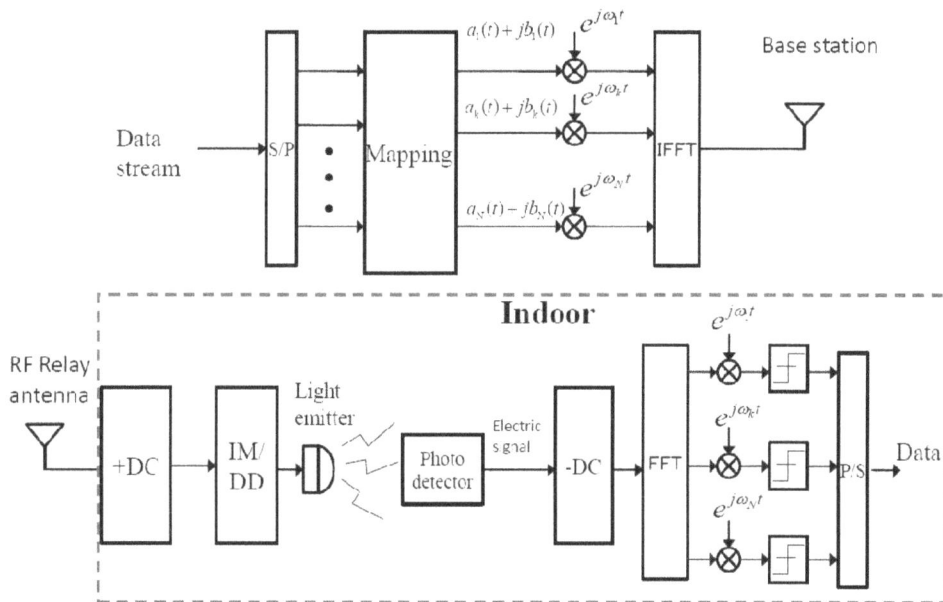

Fig. 2 Configuration of OFDM signal transmission over optical wireless link.

4.1 IM/DD optical channel

In the OFDM transmission system, the data stream is divided into N narrowband data streams, N corresponding to the subcarriers. That is to say one OFDM symbol consists of N symbols. The amplitude of the OFDM baseband signal at each sample point $t=n\Delta T$ in the effective symbol period is shown as:

$$x(n\Delta T) = \sum_{k=0}^{N-1}\{d_k e^{j2\pi\frac{nk}{N}}\} = \sum_{k=0}^{N-1}\{a_k\cos(2\pi\frac{nk}{N}) + jb_k\sin(2\pi\frac{nk}{N})\} \tag{1}$$

where $n=t/\Delta T$, a_k+jb_k is a complex symbol of kth subcarrier after mapping. The OFDM signal for N subcarriers, after up-conversion to the wireless service carrier frequency f_{RF}, can be written as:

$$S_{RF}(t) = \sum_{k=0}^{N-1}\{a_k\cos(2\pi\frac{nk}{N})\cos(2\pi f_{RF}t) + jb_k\sin(2\pi\frac{nk}{N})\sin(2\pi f_{RF}t)\} \tag{2}$$

In this paper, the signal is assumed to be intensity modulated in a linear area of the light emission element. The average absolute value of the amplitude of a_k+jb_k is assumed to be one. The optical transmission signal after fixed direct current (DC) component is added can be expressed as [18]:

$$P_t(t) = P_{opt}\{1 + m\cdot(S_{RF}(t) + n_{RF})\} \tag{3}$$

Here, P_{opt} represents the average power radiated by the light emission element. n_{RF} is an additive white Gaussian noise (AWGN) to the RF signal, and m is the modulation index. When $m = 0$, no modulation of the carrier is performed. The absolute value of mS_{RF} is smaller than 1 because the optical signal is nonnegative. If mS_{RF} is greater than 1, the carrier is actually cut off for some period of time, and unwanted harmonics are created at the transmitter output.

In this paper, Lambetian radiation intensity pattern is considered. Supposing the considered indoor OWC is performed in a relatively large space, reflection light is neglected due to its lower energy comparing to line of sight (LOS) light at the receiver side. The intensity optical channel of this system is:

$$h = \frac{1}{2\pi}\cos\theta\cos\psi\ rect(\theta/FOV)\frac{A_{OR}}{R_d^2}\delta(t - R_d/c_o) \tag{4}$$

θ is the angle between the beam central direction and the line between the transmitter and receiver. ψ is the angle between the direction vertical to the aspect of the optical detector and the line between the optical transmitter and receiver. R_d is the distance between the optical transmitter and receiver, A_{OR} is the photosensitive area of the optical detector where the incident radiation is collected and converted to an electronic signal. FOV is the angle of field of view. c_o is light speed. $\delta(t - R_d/c_o)$ is the delayed function due to the signal propagation. $rect(x)$ is the rectangular function defined as:

$$rect(x) = \begin{cases} 1\ for|x|\leq 1, \\ 0\ for|x|>1 \end{cases} \tag{5}$$

Then the received optical power P_R at the mobile terminal is given by:

$$P_R = P_t \otimes h = P_e\{1 + m \cdot (S_{RF} + n_{RF})\} \tag{6}$$

Here, P_e is the average received optical power. n_{RF} is the Additive White Gaussian Noise (AWGN) at the RF transceiver of relay system, and m is the modulation index.

Assuming that the DC term P_e can be eliminated at the output of the optical receiver, the received information signal S_{sig} coming from the RF signal is:

$$S_{sig} = S_{RF}P_e mr \tag{7}$$

Here, r represents the photodetector responsively. The average power of the information signal P_{sig} is

$$P_{sig} = P_s(P_e mr)^2 \tag{8}$$

where P_s is the average power of signal S_{RF}.

4.2 Noise model and signal to noise ratio (SNR) analysis

In this paper, we assume that the noise affecting the optical parts of our system is generated at the optical receiver. Here, Avalanche Photo Diode (APD) is assumed to be the optical receiver. Shot noise σ^2_{shot} and thermal noise σ^2_{th} are considered in our analysis. Both shot noise and thermal noise are modelled as additive white Gaussian noise (AWGN) with a double-sided power spectral density σ^2. The expression of the noise variance is given by

$$\sigma^2_{opt} = \sigma^2_{shot} + \sigma^2_{th} \tag{9}$$

$$\sigma^2_{shot} = 2qrP_eB + 2qrp_{bg}\Delta\lambda_{nb}A_{bg}B \tag{10}$$

$$\sigma^2_{th} = \frac{4kTB}{R_L} \tag{11}$$

Here, (10) includes the shot noise from the background light and the shot noise from the optical signal. Each parameter in (10) (11) are described in table 1.

For the noise of the whole process, it should include the noise generates when the RF signal is received and when the optical signal is received. The SNR from the base station to the mobile terminal can be shown as:

$$\Gamma_{all} = \frac{P_s(P_e mr)^2}{(P_e mr)^2\sigma^2_{RF} + \sigma^2_{opt}} \tag{12}$$

σ^2_{RF} is the noise generated at the relay station when radio signal is received.

If the SNR of the transceiver of the relay station is defined as $\Gamma_{ant,}$, P_s is given as

$$P_s = \sigma^2_{RF} \Gamma_{ant} \tag{13}$$

Form (12)-(13), Γ_{all} can be rewritten as:

$$\Gamma_{all} = \frac{(P_e mr)^2 \sigma^2_{RF}}{(P_e mr)^2 \sigma^2_{RF} + \sigma^2_{opt}} \Gamma_{ant} \tag{14}$$

Therefore, the ratio K_{SNR} of Γ_{ant} to Γ_{all} can be expressed as follows:

$$K_{SNR} = \frac{(P_e mr)^2 \sigma^2_{RF}}{(P_e mr)^2 \sigma^2_{RF} + \sigma^2_{opt}} \tag{15}$$

Table1. Parameters of shot noise and thermal noise

Symbol	Quantity	Value
A_{bg}	Detector area	1(cm2)
$\Delta\lambda_{nb}$	Noise-bandwidth factor	30(nm)
q	Electron charge	1.6×10-19 (C)
r	Photodetector responsively	0.53 (A/W)
p_{bg}	Background irradiance	5.8
k	Boltzmann's constant	1.374×10-23 (J/K)
T	Temperature	300(K)
R_L	Load resistance	240(Ω)
B	Filter bandwidth	1M(Hz)

4.3 Channel model in fading environment

It is necessary to consider the propagation environment of the RF signal and the indoor optical wireless signal because our system is composed of two parts that use an existing RF communication and an optical wireless communication. In our paper, assuming people indoor are moving slowly and the power of multipath light is very low, we ignore light fading. In outdoor radio communications, because the antenna for the RF wave of the relay station is set up outside, we may assume that we are having a Rayleigh fading channel or Rice fading channel. K_f is defined as the Rice factor [19]. In indoor radio communications, normally we consider that we are having a Rayleigh fading channel.

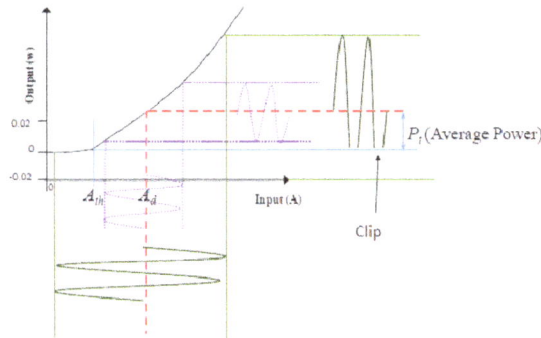

Fig. 3 Output Character of LD

5. Evaluation of characteristics of proposal method by simulation

According to the study give in the previous section, in order to know the feasibility of the proposed system and to analyze the bit error rate (BER), the RF noise ratio compared to the entire system noise, the power of the optical signal, the SNR of the RF transceiver and the fading environment, the peak clip of OFDM signal must be considered. In this chapter, the BER performance considering peak clip of OFDM signal and different radio wave environments is evaluated by simulation.

PAPR is one of the main drawbacks of OFDM system. Shown as Figure 3, when PAPR is high, peak clip will happen and influence the transmission characteristic because RF signal is intensity modulated to optical wireless signal. In this paper, we assume that the positive peak of the RF signal is completely modulated to optical signal, and the negative peak will be clipped when the threshold is exceeded. If we define the threshold current value of LD (Laser Diode) output, the bias current added to the signal, the modulation index of the intensity modulation (refer to function (3)), the amplitude of the OFDM signal as A_{th}, A_d, m, A_s, the condition that the peak clip does not happen is shown as:

$$mA_s < 1 - A_{th} / A_d \tag{16}$$

Table 2 Parameters of simulation

Number of FFT points	128
Number of sub channels: k	128
Length of Pilot	32
Modulation	QPSK
Symbol rate	250000bps
Synchronization	Perfect
Channel Estimation	Perfect

Figure 4 is the simulation result for $A_{th}/A_d = 0.33$ and $P_e = $ -15dBm. The simulation parameter is shown in table 2. From Figure 4, it can be confirmed that the BER characteristic improves as Γ_{ant} grows. Moreover, it is clearly shown that m in Equation (3) infects the transmission performance very much. With the increase of m, the BER characteristic can be separated to three regions where the BER performance relatively increases, then keep unchanged, finally deteriorates. First, as m increases, the BER performance improves since m increases the ratio of information signal power to the total optical power. However after m becomes bigger than a certain threshold, the BER performance changes slowly because m cannot change the channel impairments due to the radio noise generated at RF relay station. When m is very large, the pick clip happens and BER performance is decreased.

Figure 5 shows the BER performance versus P_e for Γ_{ant}=10dB. From this figure, we can see that the higher value of K_f, the lower values of BER can be obtained. Furthermore, when the average optical receiving power P_e is low, the value of BER changes rapidly by change P_e. When P_e is larger than -20dBm, the value of BER becomes almost independent of P_e. That is to say, the radio environment of the RF transceiver become more important than P_e to BER when P_e is larger than -20dBm.

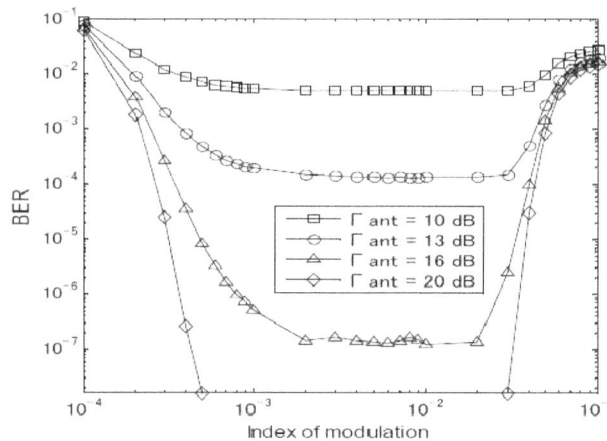

Fig.4 BER via index of modulation (P_e = -15dBm)

The transceiver of the relay station for RF signals is proposed to be located at a suitable place to make it easier to receive the preferable wave from the base station. The SNR of the RF antenna at the relay station can be much better than the SNR of the conventional RF antenna at our mobile terminal. Here, the SNR of the relay RF antenna is defined as Γ_{ant} and the SNR of the RF antenna of the mobile terminal is defined as T_m. Figure 6 shows BER performance versus received average optical power P_e and SNR for different ratio of Γ_{ant} to Γ_m in a Rayleigh fading channel when using the proposed relay system and the conventional system. From this figure, it is clear that the same as Figure 5, when P_e is bigger than a threshold, the radio environment of the RF transceiver become more important than optical link. That is to say, when the RF environment is available and the received optical power is big enough, the relay system can provide an available communication. It also can be seen that the BER performance can be improved by using higher quality RF relay antenna.

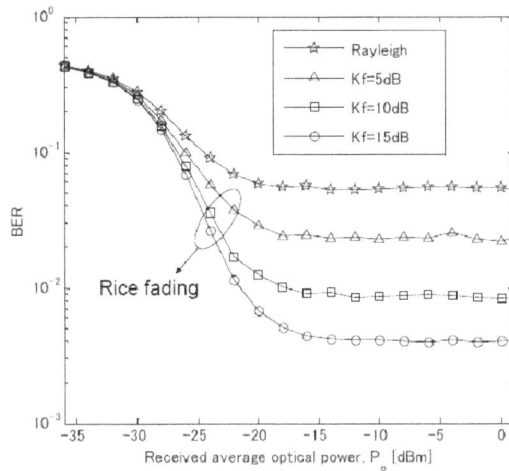

Fig.5 BER via received average optical power P_e (m=0.001, Γ_{ant}=10dB)

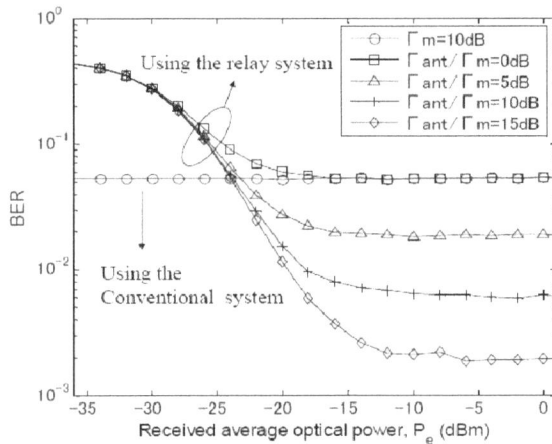

Fig.6 BER via received average optical power P_e and SNR (m=0.001, Γ_m=10dB)

6. Conclusion

In this paper we have proposed a relay system with RF subcarrier based on optical IM/DD channel. First, the advantages of using OWC have been described. Second, the concept of RoOW has been proposed. Then BER was calculated considering the possible peak clip of OFDM signal and different wave environments. Simulation result showed that the modulation index m significantly affects the peak clip of OFDM signal. It also showed that optical transmit power, radio environments such as Rayleigh fading, Rice fading are important factors which affect the transmission performance of the proposed system. It should be noted that when the received optical power is larger than -20dBm, which is a very practical power level, the value of BER becomes almost independent of the optical power and the radio environment of the RF signal become more important. That means the indoor OWC has very high transmission quality and proposed relay system is a very available system as long as the radio environment is eligible. We believe it can provide practical wireless services for RF sensitive areas.

ACKNOWLEDGEMENTS

The authors wish to thank the anonymous reviewers for their valuable suggestions.

REFERENCES

[1]   DR. Tsrmurthy and D. Siva Rama Krishna, "Analysis of cell phone usage using correlation techniques", International Journal of Wireless & Mobile Networks (IJWMN) Vol. 3, No. 2, April 2011

[2]   Eisuke hanada, Tohru Kubota, Hiroaki Shimokawa, "Permission of cellular-phone use in university hospital" 17th Autumn conference of Japanese Society of Medical Electronics and Biological Engineering, OS15-01, Oct.2003

[3]   Yoshiaki Tarusawa, Kohjiroh Ohshita, Yasunori Suzuki, Toshio Nojima, and Takeshi Toyoshima, "Experimental Estimation of EMI From Cellular Base-Station Antennas on Implantable Cardiac Pacemakers" IEEE Transactions on Electromagnetic Compatibility, VOL. 47, NO. 4, NOVEMBER 2005.

[4]   Hiroshi Takano, Yasunori Futatsugi, Shigeru Shimamoto, Kiyotaka Seki, "Studies on Optical Wireless Communications in Train and Optical Signal Propagation Model" Transmission,"IEICE Trans. Communications, Vol.J87-B, No.7, pp.950-962, July.2004

[5]   Carruthers, J.B. and J.M. Kahn, Angle diversity for nondirected wireless infrared communication. IEEE Transactions on Communications, 2000. 48(6): p. 960-969.

[6]   O'Brien, D.C., et al. High speed integrated optical wireless transceivers for in-building optical LANs. in Optical Wireless Communications III. 2000. Boston: SPIE.

[7]   Jivkova, S. and M. Kavehrad, Holographic optical receiver front end for wireless infrared indoor communications. Applied Optics, 2001. 40(17): p. 2828-35.

[8]   J.R. Barry, Wireless infrared communications, Kluwer, Boston,1994.

[9]   H. Willebrand and B. Ghuman, Free Space Optics: Enabling Optical Connectivity in Today's Networks. London, U.K.:Sams, 2002.

[10]   V. W. S. Chan, "Free-space optical communications", J. Lightwave Technol., vol. 24, no. 12, pp. 4750–4762, Dec. 2006.

[11]   Pawan Sharma and and Seema Verma, "PAPR reduction of OFDM signals Using selective mapping with TURBO code", International Journal of Wireless & Mobile Networks (IJWMN) Vol. 3, No. 4, August 2011

[12]   R. V. Nee and R. Prasad, OFDM for Wireless Multimedia Communications. Norwell, MA: Artech House, 2000.

[13]   Jayakumari.J, "MIMO-OFDM for 4G Wireless Systems", International Journal of Engineering Science and Technology, Vol. 2(7), pp. 2886-2889, 2010

[14]   Xiaodong Wang, "OFDM and its application to 4G", 14th Annual WOCC 2005. International Conference on Wireless and Optical Communications, 2005.

[15]   Hussain I., Hussain S., Khokhar, I., Iqbal, R., "OFDMA as the Technology for the Next Generation Mobile Wireless Internet" Third International Conference on Wireless and Mobile Communications (ICWMC), Guadeloupe, 2007

[16]   Roy You and J.M. Kahn, "Upper Bounding the Capacity of Optical IM/DD Channels with Multiple-Subcarrier Modulation and Fixed Bias Using Trigonometric Moment Space Method", IEEE Trans. on Information Theory, vol. 48, Issue 2, February 2002.

[17]   Gamage P.A., Nirmalathas A., Lim, C., Novak D., Waterhouse, R., "Design and Analysis of Digitized RF-Over-Fiber Links" Journal of lightwave technology, Volume: 27, Issue: 12, pp. 2052-2061, 2009

[18]   Wei Huang, Jiro Takayanagi, Tetsuo Sakanaka, Masao Nakagawa, "Atmospheric Optical Communication System Using Subcarrier PSK modulation", IEICE Transactions on Communications Vol.E76-B No.9 pp.1169-1177, Sep. 1993.

[19]   Yoshio Karasawa, Radiowave Propagation Fundamentals for Digital Mobile Communication, Japan: Corona Publishing CO.,LTD., Tokyo, August , 2006 (in Japanese)

15

ENFORCING END-TO-END PROPORTIONAL FAIRNESS WITH BOUNDED BUFFER OVERFLOW PROBABILITIES IN AD-HOC WIRELESS NETWORKS

Nikhil Singh[1] and Ramavarapu Sreenivas[2]

[1]Yahoo! Labs, Champaign, IL 61820, USA

[2]Coordinated Science Laboratory & Industrial and Enterprise Systems Engineering, University of Illinois at Urbana-Champaign, Urbana, IL 61820

ABSTRACT

In this paper, we present a distributed flow-based access scheme for slotted-time protocols, that provides proportional fairness in ad-hoc wireless networks under constraints on the buffer overflow probabilities at each node. The proposed scheme requires local information exchange at the link-layer and end-to-end information exchange at the transport-layer, and is cast as a nonlinear program. A medium access control protocol is said to be proportionally fair with respect to individual end-to-end flows in a network, if the product of the end-to-end flow rates is maximized. A key contribution of this work lies in the construction of a distributed dual approach that comes with low computational overhead. We discuss the convergence properties of the proposed scheme and present simulation results to support our conclusions.

KEYWORDS

Wireless LAN, Access protocols, Resource management.

1. INTRODUCTION

In this paper we consider an ad-hoc wireless network [1] that carries several flows between various source-destination pairs under a slotted-time medium access control (MAC) protocol. Specifically, we are interested in a distributed scheme for the assignment of the network's resources among flows, which is fair in terms of end-to-end flow rates. We assume that each node in the network has a finite buffer assigned to each flow routed through it. In addition to the objective of fairness, we are also interested in ensuring that the buffer overflow probability at each node does not exceed a pre-determined value.

The literature contains several references to fairness and its impact on the network performance. It has been observed by many researchers that the contention control mechanism used in 802.11-MAC [2] can be inefficient [3]. In [4], [5] a list of modifications is presented, that eliminates the unfairness commonly seen in the 802.11-MAC. The literature also contains a large volume of references (cf. [6], [7], [8], for example) where it is assumed that each network flow/link is associated with a concave utility function that could be maximized. In particular, for proportional fairness, it is assumed that the utility function has the form of *log x*, where *x* denotes the flow rate

[6]. It is of interest to schedule individual transmissions on the links so as to maximize the sum of the utilities of the consumers. To achieve fairness, the schemes outlined in the above-mentioned references use a penalty function that is updated by some form of feedback from the network. Using an appropriately defined cost that is implicitly dependent on the requested rates of each node within a neighbourhood, the penalty is typically the total cost of all nodes in the network. A node maximizes (its view of) a common performance function, given by the difference between the total utility and the penalty. An overview of network resource allocation through utility maximization is presented in [9].

In [10], the authors have addressed the problem of providing proportional fairness by considering joint optimization at both transport and link layers. Two algorithms are proposed for solving the problem in a distributed manner that converges to the globally optimal solutions. These results, generalized in [11], are based on the dual and the primal algorithms in convex optimization and need end-to-end feedback information to update variables maintained at the nodes. The algorithms presented in [10], [11] are oblivious of the queue dynamics of the network, which may increase delays and packet loss. Although our work is closely related to [10], [11], the problem formulation and the proposed solution differ significantly.

In [12], the solution approach uses a class of *queue backpressure random access algorithms* (QBRA), where the actual queue-lengths of the flows are used to determine any node's channel access probabilities. In this distributed algorithm, a node uses the queue-length information in a close neighbourhood to determine its channel access probability to achieve proportionally fair rates and queue stability. This scheme has the advantage that no optimization needs to be performedand nodes can achieve proportional fairness just by exchanging the queue information in the local neighbourhood. However, the frequency of exchange of this information plays a vital role in determining the performance of this algorithm. In optimization-based schemes, once the flow rates have converged to the optimum, the frequency of information exchange does not play a significant role until the network topology, or the number of flows in the network, change.

In a different approach, several policies have recently been proposed for achieving rates close to the maximum throughput region through dynamic link scheduling [13], [14], [15], [16]. These scheduling algorithms use maximal matchings in every time slot using local contention algorithms and achieve near maximal schedules. Some policies also guarantee fairness of rate allocation among different sessions.

Quality of Service(QoS) is an important issue in ad-hoc wireless networks. Service guarantees can be provided for delays, packet loss, jitter and throughput based on the application requirements. Our approach in this work is to combine the QoS guarantee in addition to providing proportional fairness. Our main contributions are as follows:

1. We derive an expression for the buffer overflow probabilities for discrete-time queues. This derivation uses the fact that there cannot be simultaneous arrivals and departures at a node within the same slot in Aloha-type networks that do not have packet capture mechanisms.
2. Using the expression for buffer overflow probabilities mentioned above, we show that an upper bound on the buffer overflow probability translates to an upper bound on the utilization or load, which can then be used as constraints in an appropriately posed convex minimization problem under convex constraints. This is a reformulation of the proportionally fair end-to-end rate allocation problem. A distributed dual approach is then used to solve this convex minimization problem using an appropriate Lagrangianfunction. The dual problem is solved using a projected gradient method.
3. Finally, after making some observations about the distributed implementation of the above-mentioned dual scheme, we present simulation results showing the satisfactory

performance of our proposed algorithm in terms of fairness and QoS.

The rest of the paper is organized as follows. Section 2 presents the network model that is used in the rest of the paper. We then formulate the rate control problem as a convex optimization instance with bounds on the buffer overflow probabilities at each node. In section 3, we discuss the dual-based solution approach and present a distributed implementation to achieve flow-based proportional fairness. The convergence of this algorithm to the unique global optimum is established. Section 4 contains the details of the experimental results verifying the optimality of the proposed scheme. Conclusions are provided in section 5.

2. PROBLEM FORMULATION

2.1. Wireless Network Model

We assume the following:

1. Time is divided into slots of equal duration.
2. A successful transmission in a time-slot implies collision free data transmission in that slot.
3. The transmitting nodes always have data packets to transmit (i.e. we do not consider □the arrival rates of packets for different flows, and assume that all flows have packets to transmit at all times).
4. Nodes cannot transmit and receive packets at the same time.
5. The receipt of more than one packet within the same time-slot will result in a collision.
6. Nodes in the network have a buffer of fixed size assigned to each flow that is routed through it.
7. We also assume there is a unique route for each flow within the network (which would be the case if we used DSDV [17] as the routing protocol, for example).

Additionally, we only consider unicast flows for our derivations.

An ad-hoc wireless network carrying a collection of flows, is represented as an undirected graph $G = (V, E)$, where V represents the set of *nodes*, and $E \subseteq V \times V$ is a symmetric relationship (i.e. $(i, j) \in E \Leftrightarrow (j, i) \in E$), that represents the set of bidirectional *links*. We assume all links of the network have the same capacity, which is normalized to unity. The 1-hop neighbourhood of node $i \in V$ is represented by the symbol $N(i)$. When a node i communicates with a node $j \in N(i)$, we can represent it as an appropriate orientation of the link *(i, j)* in E, where i is the origin and j is the terminus. The context in which $(i, j) \in E$ is used should indicate if it is to be interpreted as a directed edge with i as origin and j as terminus. The set of flows, using a link $(i, j) \in E$ with $i(j)$ as origin (terminus), is denoted by $\mathcal{F}(i, j)$.

When node i intends to transmit data to node $j \in N(i)$ for the l-th flow ($l \in \mathcal{F}(i, j)$), it would transmit data in the appropriate time-slot with probability $p_{i,j,l}$. $\mathcal{P}_{i,j} = \sum_{l \in \mathcal{F}(i,j)} p_{i,j,l}$, denotes the probability that node i transmits data to node j, and $\mathcal{P}_i = \sum_{j \in V} \mathcal{P}_{i,j}$, denotes the probability that node i will be transmitting to some node in its 1-hop neighbourhood for some flow. The probabilities $p_{i,j,l}$'s should be chosen such that $\mathcal{P}_i$ is not greater than unity for any node $i \in V$.

2.2. Link Success Probability Expression

The probability of successful data transmission over link *(i, j)* □E for flow l □$\mathcal{F}$ *(i, j)*, denoted by

$S_{i,j,l}$, is given by the expression

$$S_{i,j,l} = p_{i,j,l}\left(1 - \sum_{(j,m)\in E, n\in \mathcal{F}(j,m)} p_{j,m,n}\right) \prod_{o\in N(j)-\{i\}}\left(1 - \sum_{(o,p)\in E, q\in \mathcal{F}(o,p)} p_{o,p,q}\right) \quad (1)$$

This is also the rate or the attainable throughput of flow l over link (i, j).

2.3. Problem Statement

Consider an ad-hoc wireless network where there are r flows in the network. Each flow has a utility function associated with it, whose value is determined by the logarithm of the flow rate. The objective is to maximize the sum of the logarithms of the flow-rates under the operational constraints outlined below. We denote the logarithm of the rate of the l-th flow as f_l. The end-to-end proportionally fair flow control problem can be stated as

$$\max_{p_{i,j,l}} \sum_l f_l \quad (2)$$

where $(i, j) \square E$ and $l \square \{1, 2, \ldots, r\}$, subject to additional constraints.

Let us assume that the l-th flow ($1 \leq l \leq r$) spans over k_l links. We use the notation $\langle l, q \rangle \in E$ to denote the l-th-flow's q-th-link, where is $q \in \{1, 2, \cdots, k_l\}$ is indexed in ascending order starting from the source and terminating at the destination. Thus, $\langle l, q \rangle = (i, j)$ implies the l-th-flow's q-th-link from the source has i as the source node and j as the destination node. If $\langle l, q \rangle = (i, j) \in E$ then we use the notation $S_{l,q}$ to denote $S_{i,j,l}$. The logarithm of the rate of l-th flow over link $\langle l, q \rangle$ is represented as $f_{l,q}$.

Let $\mathbf{p} = (p_{l,q}, 1 \leq l \leq r, 1 \leq q \leq k_l, \langle l, q \rangle \in E)$ bethevectorofaccessprobabilitiesofallthe

flowsovereachlinkinthenetworkand $\hat{\mathbf{f}} = \left(f_{l,q}, 1 \leq l \leq r, 1 \leq q \leq k_l, \langle l, q \rangle \in E\right)$ thevectorof   the logarithm of link rates of all flows.

In the case of multi-hop wireless networks, the rate of any flow is the same as the rate of the bottleneck link in that flow. The logarithm of the rate of the l-th flow is $min\{f_{l,q} : 1 \leq q \leq k_l\}$. Hence, the problem can be stated as $\max\limits_{p_{l,q}} \sum\limits_l \min\{f_{l,q}, 1 \leq q \leq k_l\}$, subject to capacity constraints, and additional constraints on the buffer overflow probabilities which is addressed in the nextsubsection.

2.4. Buffer Overflow Probability of a Tandem of Discrete-Time Queues

The results in reference [18] can be paraphrased as follows – for a discrete-time queue of capacity M, with a packet arrival probability p_a, and a probability p_d ($p_d > p_a$) of a packet departure from a non-empty buffer, the probability of seeing i-many packets at any time-instant in the buffer in steady state is given by the expression

$$\frac{1-\left(\dfrac{p_a}{p_d}\right)}{1-\left(\dfrac{p_a}{p_d}\right)^{M+1}}\left(\dfrac{p_a}{p_d}\right)^{i} \qquad (3)$$

Using the time-reversibility of the underlying Markov-chain, and the mutual independence of the simultaneous states of the buffers, reference [18] also establishes that the joint stationary state probability of a tandem of discrete-time queues is the product of the distributions of each queue taken independently with an arrival probability of pa, which is the probability of packet arrival into the first queue. This is essentially the discrete-time analogue of Jackson's result [19] involving tandems of M/M/1 queues. The key points of divergence between reference [18] and the present paper are presented below.

It should be noted that unlike the model assumed in reference [18], where arrival and departure events are permitted to occur concurrently, interference constraints in wireless networks do not permit the occurrence of certain simultaneous events. For instance, as a node cannot transmit and receive information at the same time, the simultaneous occurrence of an arrival and a departure from the discrete-time queue at the node cannot be permitted. Secondary interference constraints place additional restrictions on the set of simultaneous events that can occur among neighbouring nodes. Even when there are no restrictions on simultaneous events, reference [20] notes that it is cumbersome to use balance equations to arrive at an appropriate expression for the joint stationary probability for tandems of discrete-time queues. For situations where there are restrictions on the nature of concurrent events that can occur in a tandem of queues, such as those that model wireless networks, the joint stationary state probability of a tandem of discrete-time queues is not guaranteed to have the product-form of reference [18]. This notwithstanding, it is possible to characterize the marginal probability distribution of each queue in the tandem.

We first note that the analysis of reference [18] (cf. equations 1, 2 and the subsequent discussion of time-reversibility) applies *mutatis mutandis* to the case when utmost one packet is permitted to arrive, or depart from a single discrete-time queue of size M, along with the restriction that a simultaneous arrival and departure of a packet from the queue is not permitted. The probability of seeing i-many packets in the buffer at any time-instant in this restricted discrete-time queue is also given by equation 3. The probability of the queue of size M is non-empty is given by the expression

$$\frac{1-\left(\dfrac{p_a}{p_d}\right)^{M}}{1-\left(\dfrac{p_a}{p_d}\right)^{M+1}}\left(\dfrac{p_a}{p_d}\right)$$

and since the probability of a packet departure from a non-empty queue is p_d, the probability of a packet-departure from the discrete-time queue is given by

$$\frac{1-\left(\dfrac{p_a}{p_d}\right)^{M}}{1-\left(\dfrac{p_a}{p_d}\right)^{M+1}}\left(\dfrac{p_a}{p_d}\right)\times p_d < p_a.$$

It is not hard to see that if $M = \infty$, then the probability of a packet-departure from the discrete-time queue is exactly equal to the probability of packet-arrival into the queue. For bounded queues ($M < \infty$) the output process of the queue is geometrically distributed with a parameter that is no greater than the input parameter p_a. Additionally, there can be no more than M-many consecutive departures, or, M-many consecutive arrivals to the discrete-time queue due to the bound on the buffer-size. We assume packets that arrive into a full-queue get dropped. This observation holds for a tandem of discrete-time queues. That is, the output process of each queue is geometrically distributed with a parameter that is no greater than that of the input to the first queue (i.e. p_a). This observation is used in establishing a bound on the buffer-overflow probabilities at each queue in a tandem of discrete-time queues in the following theorem.

Theorem 1.1: Consider a tandem of n discrete-time queues, each with buffer-size M, whereat any discrete-time instant the probability of a packet-arrival into the first queue is p_a, andthe probability of a packet-departure from the i-th, non-empty queue is p_{di}, ($i = 1, 2, ..., n$). If

$$p_{dj} = \min_{i=1,\Box,n} \{p_{di}\}, \text{ and } \frac{p_a}{p_{dj}} < \frac{M}{M+1}, \text{then, the probability of seeing } M \text{ packets in the } i\text{-th queue } (i=$$

1, ...,n) is no greater than

$$\left(\frac{1 - \dfrac{p_a}{p_{dj}}}{1 - \left(\dfrac{p_a}{p_{dj}}\right)^{M+1}} \right) \left(\frac{p_a}{p_{dj}} \right)^M$$

Proof: Suppose $\rho = \dfrac{p_a}{p_{dj}}$, we first note that the expression $\left(\dfrac{1-\rho}{1-\rho^{M+1}} \right)\rho^M$, increases

monotonically with respect to ρ if $\rho \leq \dfrac{M}{M+1}$. Let p_{ai} be the probability of a packet arrival into the

i-th queue, weknow$p_{ai} \leq p_a$.If $\rho_i = \dfrac{p_{ai}}{p_{di}}$,since$p_{di} \geq p_{dj}$,itfollowsthat $\rho_i \leq \rho < \dfrac{M}{M+1}$. Theobservation

follows directly from the monotonicity property mentioned above.■

A direct consequence of theorem 1.1 is that if we are able to pick a p_a such that

$$\frac{p_a}{p_{dj}} < \left[\frac{\beta}{1+\beta} \right]^{1/M},$$

then the buffer overflow probability at the i-th queue in the tandem of discrete-time queues will be no higher than β at all queues. In the next section, this observation is used in a convex programming solution to the problem of enforcing proportional fairness in the presence of constraints on the buffer overflow probabilities.

2.5. Problem Formulation with Buffer Overflow and Capacity Constraints

Let us assume the loss rate bounds for the l-thflow translates to each node along the flowsustaining a traffic intensity (ratio of arrival probability and departure probability at a node) nomore than $\rho_l \left(= \dfrac{p_a}{p_{dj}} \right)$.

Also, each link-rate in the network cannot exceed the capacity of that link given by (1). Since the logarithmic function is strictly increasing, each link constraint can be re-written as

$$f_{l,q} \leq \log(\mathcal{S}_{l,q}) \qquad (5)$$

Each link constraint (5) forms a convex set over $(f_{l,q}, \mathbf{p})$. We also assume that there is a minimum achievable data-rate for each flow, i.e., $\exists \varepsilon, s.t. \varepsilon \leq f_{l,q}, \forall l, q (1 \leq l \leq r, 1 \leq q \leq k_l)$. Also, we assume that all the flows in the network have a maximum achievable data-rate i.e., $\exists \delta, s.t. f_{l,q} \leq \delta, \forall l, q (1 \leq l \leq r, 1 \leq q \leq k_l)$,(toaccommodatenetworkcontrol trafficlike routing messages, for example). We define the feasible set of access probabilities as,

$$\tilde{\mathcal{P}} = \left\{ \mathbf{p} : \sum_{\{l:\langle l,q \rangle = (i,j), j \in N(i)\}} p_{l,q} \leq 1, e^{\varepsilon} \leq p_{l,q} \leq e^{\delta}, (i,j) \in E, l \in \mathcal{F}(i,j) \right\}.$$

Also, we define the *QoS region* as a set of vectors as defined by

$$\mathcal{G} = \left\{ \hat{\mathbf{f}} : \varepsilon \leq f_{l,1}, f_{l,q} \leq \delta, f_{l,1} \leq f_{l,q} + \delta_l, 2 \leq q \leq k_l \right\},$$

where$\delta_l = \log \rho_l$. The overall optimization problem can now be stated as:

$$\mathbf{V}: \quad \max : \sum_l \min \left\{ f_{l,q} : 1 \leq q \leq k_l \right\}, \qquad (6)$$

$$f_{l,q} \leq \log(\mathcal{S}_{l,q}), \forall \langle l, q \rangle \in E,$$

$$\mathbf{p} \in \tilde{\mathcal{P}}, \hat{\mathbf{f}} \in \mathcal{G}.$$

From the constraint imposed by the QoS region, we observe that for any feasible solution to **V**, the first link will always have the lowest rate and hence it will be the bottleneck. Therefore for any feasible solution, the rate of any flow l, is same as fl,1. We replace $f_{l,1}$ by f_l, and define the feasiblesetofflowratesas $\tilde{\mathcal{F}} = \{ \mathbf{f} : \varepsilon \leq f_l \leq \delta, \forall l \}$, where,$\mathbf{f} = (f_l, 1 \leq l \leq r)$,wecanrewrite**V** as the following convex optimization problem,

$$\mathbf{U}: \quad \min \sum_l -f_l, \qquad (7)$$

$$f_l \leq \log(\mathcal{S}_{l,1}), \forall \langle l, 1 \rangle \in E,$$

$$f_l \leq \log(\mathcal{S}_{l,q}) + \delta_l, \forall \langle l, q \rangle \in E, 2 \leq q \leq k_l,$$

$$\mathbf{p} \in \tilde{\mathcal{P}}, \hat{\mathbf{f}} \in \tilde{\mathcal{F}}.$$

3. SOLUTION APPROACH

3.1. Dual-based Algorithm

We can write the Lagrangian function for the problem stated in (7) as,

$$\mathcal{L}(\mathbf{f},\mathbf{p},\lambda)=\sum_l -f_l +\sum_{l,1}\lambda_{l,1}(f_l -\log(S_{l,1}))+\sum_{l,q(2\leq q\leq k_l)}\lambda_{l,q}(f_l -\log(S_{l,q})-\delta_l). \tag{8}$$

Let us denote $\Lambda=(\lambda_{l,q}:\forall l,1\leq q\leq k_l)$ as a vector of Lagrange multipliers. As the Slater constraint qualification is satisfied by the convex program given by (7), convex duality implies that at the optimum Λ^*, the corresponding $\mathbf{f}$, $\mathbf{p}$ are the solutions to the primal problem [21]. The dual problem can be solved using the gradient projection method similar to the scheme used in [22]. Note that the Lagrangian is separable in terms of the probabilities $\mathbf{p}$ and the logarithm of the rates $\mathbf{f}$. The dual function can be stated as:

$$Q(\Lambda)=\inf_{\mathbf{f}\in\tilde{\mathcal{F}},\mathbf{p}\in\tilde{\mathcal{P}}}\mathcal{L}(\mathbf{f},\,\mathbf{p},\,\Lambda). \tag{9}$$

The following proposition is significant for obtaining the distributed solution for the non-linear program given by (7).

Proposition 2.1: For a given $\Lambda(\lambda_{l,q}\neq 0,\forall\langle l,q\rangle\in E)$, the solution to $\inf_{\mathbf{f}\in\tilde{\mathcal{F}},\mathbf{p}\in\tilde{\mathcal{P}}}\mathcal{L}(\mathbf{f},\,\mathbf{p},\,\Lambda)$ is given by:

$$p_{l,q}=\max\left\{\min\left\{\frac{\lambda_{l,q}}{\sum_{\langle y,z\rangle=(i,k),k\in N(i)}\lambda_{y,z}+\sum_{\langle y,z\rangle=(k,i),k\in N(i)}\lambda_{y,z}+\sum_{v\in N(i)}\sum_{\langle y,z\rangle=(k,v),k\in N(v)-\{i\}}\lambda_{y,z}},e^\delta\right\},e^\epsilon\right\}, \tag{10}$$

and,

$$f_l=\begin{cases}\varepsilon & \text{if }\sum_q\lambda_{l,q}\geq 1\\ \delta_l & \text{if }\sum_q\lambda_{l,q}<1\end{cases} \tag{11}$$

Proof: Since the Lagrangian is convex with respect to $\mathbf{p}$, the unconstrained value of $\mathbf{p}$ that yields the infimum of the Lagrangian is obtained by taking its derivative with respect to $\mathbf{p}$ and equating it to zero. This results in the expression involving λ-terms in equation 10.

The denominator of this expression is essentially the sum of three terms. The first term is the sum of the Lagrange multipliers associated with all outgoing flows from node i. The second is the sum of the Lagrange multipliers associated with all incoming flows to node i. Finally, the third term is the sum of the Lagrange multipliers associated with all incoming flows to nodes in the one-hop neighborhood of node i (excluding the flows incoming from node i). It is not hard to show that this expression satisfies the constraint $0\leq p_{l,q}\leq 1$ and $0\leq P_i\leq 1$.

The remaining terms, and the structure of equation 10, are a direct consequence of the upper and lower-bounds on $p_{l,q}$ ($e^{\varepsilon} \leq p_{l,q} \leq e^{\delta}$).

Also, for any l, the coefficient of f_l in the Lagrangian, is given by $(-1 + \sum_q \lambda_{l,q})$. When $\sum_q \lambda_{l,q} \neq 1$, the infimum of Lagrangian with respect to f_l is either the lower or the upper bound of f_l, depending on the sign of its coefficient, which gives (11). When $\sum_q \lambda_{l,q} = 1, f_l$ can take anyvalue between $[\varepsilon, \delta]$. ∎

The dual problem

$$\text{maximize: } Q(\Lambda) \qquad (12)$$

$$\text{subject to: } \Lambda \in [\eta, \infty) \times [\eta, \infty) \times \square \times [\eta, \infty)$$

whereη is a small number close (but not equal) to zero, can now be solved using the subgradient projection method, where the Lagrange multipliers are adjusted in the direction of thesubgradient

$$\lambda_{l,q}^{n+1} = \left[\lambda_{l,q}^n + \alpha^n \frac{\partial Q(\Lambda^n)}{\partial \lambda_{l,q}} \right]^+ \qquad (13)$$

where$\left[z_{\langle l,q \rangle} \right]^+ = \max\{\eta, z_{\langle l,q \rangle}\}$ and $\frac{\partial x}{\partial y}$ denotes the subgradient of x with respect to y. The variableα^n is the step size at the n^{th}iteration that can either be a constant, or, diminishing step sizethat satisfies the requirements

$$\lim_{n \to \infty} \alpha^n = 0, \sum_{n=1}^{\infty} \alpha^n = \infty,$$

and the subgradient is given by,

$$\frac{\partial Q(\Lambda^n)}{\partial \lambda_{l,1}} = (f_l^n - \log(\mathcal{S}_{l,1}^n)), \qquad (14)$$

$$\frac{\partial Q(\Lambda^n)}{\partial \lambda_{l,q}} = (f_l^n - \log(\mathcal{S}_{l,q}^n) - \delta_l), 2 \leq q \leq k_l, \qquad (15)$$

wheref_l and $\lambda_{l,q}$ are obtained from equations (10) and (11).

3.2. Convergence of the Proposed Scheme

Since $\mathbf{p} \in \tilde{\mathcal{P}}$, the norm of subgradients of the dual function given by (9), is bounded. Let Λ^*be the solution to (12) for $\eta = 0$. Also let $\hat{\Lambda}_\eta$be the global maximum of (12) for $\eta > 0$.Let $\{\Lambda_\eta^n\}$be the sequence generated by the subgradient method given by (13) for $\eta > 0$.

Lemma 2.2: Let$\left\| g(Q(\Lambda_\eta^n)) \right\|_2 \leq G, \forall n \geq 0$. For every $\eta > 0$, $\exists \hat{\Lambda}_\eta$, such that

$$\left\| Q(\Lambda^*) - Q(\hat{\Lambda}_\eta) \right\|_2 \le G\eta.$$

The proof follows directly from the concavity of the dual function, and the property of the subgradients [23].

Theorem 2.3: For every $\eta > 0$, every limit point of the sequence of $\{\Lambda_\eta^n\}$ obtained using the diminishing step size, is the global maximum of (12).

Proof: Chapter 2 of reference [23], presents a proof of convergence of approaches that use subgradient method involving diminishing step sizes, which can easily be extended to projected subgradientmethod for maximization of concave function over a convex set. ∎

Theorem 2.4: For every $\eta > 0$, using the constant step size $\alpha^n = \alpha$, as $n \to \infty$,

$$Q(\hat{\Lambda}_\eta) - \liminf_{n\to\infty}(Q(\overline{\Lambda}_\eta)) \le \frac{G\alpha^2}{2}, \text{ where } \overline{\Lambda}_\eta = \frac{1}{n+1}\sum_{i=0}^{n}\Lambda_\eta^i.$$

The proof of convergence of projected subgradient method with constant step size,

underassumption of bounded subgradients, is presented in the appendix for completeness.

3.3. Implementation of the Dual-Based Algorithm

The dual-based algorithm for end-to-end proportionally fair rate allocation under buffer overflow constraints in random access wireless networks can be summarized as follows:

1. Initialize the iteration count n to zero. If $\langle l, q \rangle = (i, j)$ for some flow l, node i chooses an initial value of $\lambda_{l,q}^0$ such that $0 < \lambda_{l,q}^0 < 1$.
2. Node $i$passes the value of $\lambda_{l,q}^n$ to the source of the l-thflow. The logarithm of the rates (f_l) are then computed by sources using (11) in *O(1)* time.
3. Every node that the l-thflow is routed through, obtains the value of f_l^n from the source.
4. After obtaining the $\lambda_{l,q}^n$-values from nodes within a 2-hop neighborhood, each node computes the access probability values $(p_{l,q})$ according to (10).
5. Each node increments the value of n and computes $\lambda_{l,q}^{n+1}$ by the gradient projection algorithmgiven by (13) in *O(1)* time.
6. Steps 2, 3, 4 and 5 are repeated till an appropriate stopping condition is satisfied (based on heuristics or some other criteria, see discussion below).

When flows can arrive and depart in the network, constant step size is the preferred optionand in this case there is no stopping criteria, i.e. the nodes continue to run the optimization algorithm without termination. The access probabilities are updated periodically and the source of each flow sets the flow rate as

$$\min\left\{\log(S_{l,1}), \min_{2\le q\le k_l}(\log(S_{l,q})-\delta_l)\right\}, \qquad (16)$$

to maintain the flow rates in the QoS region.

4. PERFORMANCE EVALUATION

For our simulation comparisons, we consider the example shown in figure 1, from references [10], [12]. The nodes are labeled from 1 to 6. The interference model is that each node interferes with the reception at its one-hop neighbors. For example nodes 1 and 3 cause interference at node 2; nodes 6, 5, 2 and 4 cause interference at node 3. Three end-to-end flows, namely *flow₁*, *flow₂*, and *flow₃* are setup in this network. The source, the sinks, and the path of three flows are shown in table I.

We suppose each flow can tolerate a loss of 45 in every 100,000 packets. Additionally, we suppose each node has a buffer that can store 50 packets for each flow that is routed through it.

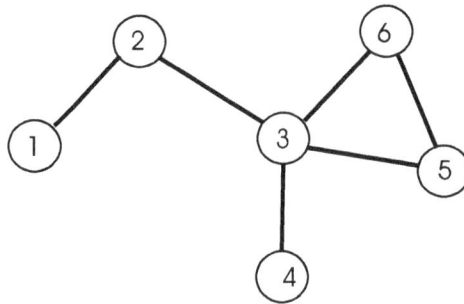

Fig. 1. An ad-hoc wireless network.

Flow	Links (Source, Sink) on the path	Observed flow rates in Matlab
$flow_1$	$\langle 1,1 \rangle = (6,5), \langle 1,2 \rangle = (5,3), \langle 1,3 \rangle = (3,2), \langle 1,4 \rangle = (2,1)$	0.0462
$flow_2$	$\langle 2,1 \rangle = (6,3), \langle 2,2 \rangle = (3,4)$	0.1123
$flow_3$	$\langle 3,1 \rangle = (1,2), \langle 3,2 \rangle = (2,3), \langle 3,3 \rangle = (3,4)$	0.0799

TABLE I: Path of the flows and observed flow rates in a MATLAB simulation of the network shown in figure 1.

This translates to a value of $\rho = 0.86$. For $\rho_1 = \rho_2 = \rho_3 = 0.86$, the globally optimal solutions to the problem defined in (7) was computed using the *fmincon* function in MATLAB. This solution and the solutions given by the dual-based algorithm are presented in table II. More detailed experimentation results can be found in [24], which has been excluded from this paper due to page limitations.

We have used two approaches to the dual-based algorithm outlined earlier. In the first approach we use a constant step size $\alpha^n = 5 \times 10^{-4}$ (cf. (13)), and the logarithm of the minimal achievable rate, ε was set to be -10. Figure 2(a) shows how the flow rates (cf. (16)), converge when the dual-based algorithm with fixed step size is used.

The second approach involves the use diminishing step size. In this case the step size at the nth iteration $\alpha^n = 1/n$. The value of ε is set to be -10. Figure 2(b) illustrates the convergence of the flow rates (cf. (16)), when the dual-based algorithm with diminishing step sizes is used.

If $\rho_1 = \rho_2 = \rho_3 = 1$, we get the optimal solution of $U^* = -7.4897$ using MATLAB's *fmincon* function. This is higher than when $\rho_1 = \rho_2 = \rho_3 = 0.86$, but the buffer overflow will be significantly higher. To demonstrate this, we simulated the network in figure 1 in

MATLAB, using access probabilities obtained for $\rho_1 = \rho_2 = \rho_3 = 1$ and $\rho_1 = \rho_2 = \rho_3 = 0.86$, and ran the simulation for a duration of 5×10^4 time slots, where each node in the network has infinite length buffers (i.e. no packets are dropped in the simulation). For a random instance of the simulation, we plot the queue-lengths as a function of time, for flow 1 at node 5. In the plots, unit of time is a single time-slot of fixed duration. The flow rates observed for $\rho_1 = \rho_2 = \rho_3 = 0.86$ are presented in table I.

Figure 3 demonstrates the queueing performance of our algorithm at node 5. Case I is the plot of queue-length as a function of time when we use the optimal access probabilities without considering buffer overflow constraints i.e. $\rho_1 = \rho_2 = \rho_3 = 1$. Case II shows how the queue- length varies as a function of time, when the optimal access probabilities obtained by setting $\rho_1 = \rho_2 = \rho_3 = 0.86$ is used. We can observe from the plots that in case I, the queue is unstable, whereas, case II shows the queue at node 5 is stable. If a buffer size of 50 was used, then the fraction of packets transmitted that are lost, in case I will be much higher than as compared to case II.

Variables	$p_{6,5,1}$	$p_{5,3,1}$	$p_{3,2,1}$	$p_{2,1,1}$
optimum solutions	0.0881	0.2185	0.1028	0.0657
dual-based solutions with constant step sizes	0.0892	0.2165	0.1078	0.0675
dual-based solutions with diminishing step sizes	0.0882	0.2191	0.1032	0.0661
Variables	$p_{6,3,2}$	$p_{3,4,2}$		
optimum solutions	0.3388	0.1329		
dual-based solutions with constant step sizes	0.3353	0.1419		
dual-based solutions with diminishing step sizes	0.3377	0.1333		
Variables	$p_{1,2,3}$	$p_{2,3,3}$	$p_{3,4,3}$	
optimum solutions	0.1776	0.2949	0.0892	
dual-based solutions with constant step sizes	0.1875	0.2929	0.0903	
dual-based solutions with diminishing step sizes	0.1761	0.2949	0.0893	
Variables	f_1	f_2	f_3	U^*
optimum solutions	0.0465	0.1143	0.0767	-7.8051
dual-based solutions with constant step sizes	0.0461	0.1109	0.0792	-7.8118
dual-based solutions with diminishing step sizes	0.0464	0.1136	0.0759	-7.8239

TABLE II: □The Optimal Results and the Solution given by the Distributed Algorithm

(a) Convergence of flow rates with constant step sizes.

(b) Convergence of flow rates with diminishing step sizes.

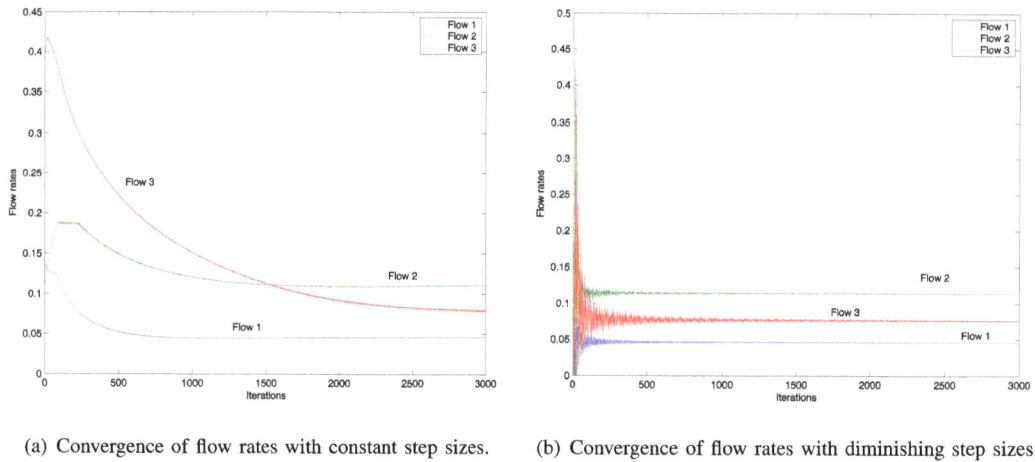

Fig. 2. The convergence of flow rates when dual-based algorithm is used.

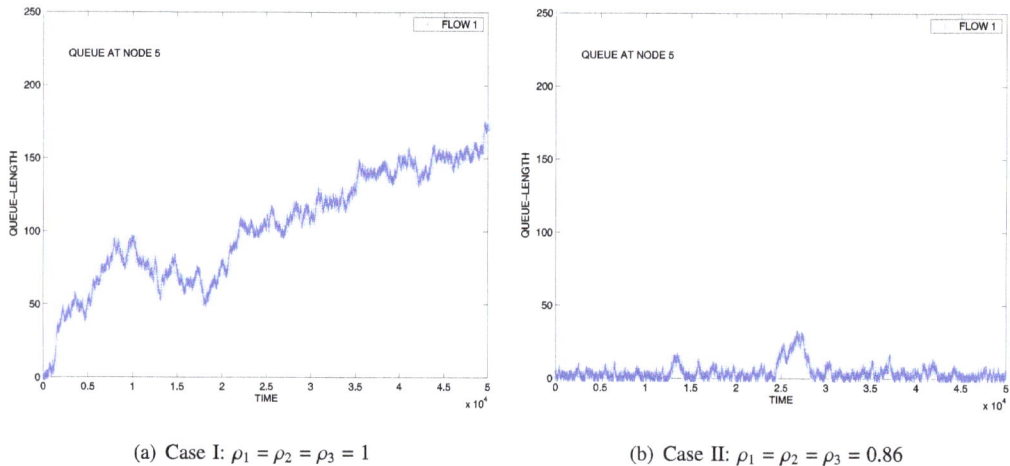

(a) Case I: $\rho_1 = \rho_2 = \rho_3 = 1$

(b) Case II: $\rho_1 = \rho_2 = \rho_3 = 0.86$

Fig. 3. The queue-length at Node 5.

5. CONCLUSION

In this paper, we proposed a distributed scheme for providing end-to-end proportionally fair flow rates in a slotted-time, multi-hop, random access network with a general network topology, with bounds on the buffer overflow probabilities at each node. After noting that each flow in the network can be viewed as a tandem of discrete-time queues, we converted the constraints on buffer overflow probabilities into appropriate constraints on the link rates, which permitted the reformulation of the original problem into an appropriately posed convex minimization problem under convex constraints. We solved this problem using an appropriately constructed Lagrange function, and discuss its convergence properties. After presenting aspects of distributed implementation of this dual-based approach, we verified the correctness of the approach using an example from the literature.

APPENDIX A: PROOF OF CONVERGENCE OF PROJECTED SUBGRADIENT METHODS WITH CONSTANT STEP SIZE

Suppose $f : \mathcal{R}^n \to \mathcal{R}$ is a concave function defined over a convex set C, having a non-emptyset of maximum points M^*. To maximize f, the projected subgradient method uses the iteration $x_{k+1} = [x_k + h_{k+1}g(x_k)]^+$, where x_k is the k-th iterate, $g(x_k)$ is the subgradient of f at x_k and h_{k+1} is the step size, and for constant step size we have $h_k = h, \forall k$. We assume that norm of subgradients of f is bounded and therefore, $\|g(x_k)\|_2 \le G, \forall k$. Also, we define at the k-th iterate, $\bar{x} = \dfrac{1}{k+1}\sum_{i=0}^{k}x_i$.

Theorem A.1: For any $x^* \square M^*$, as $k \to \infty$ one can find a $\hat{x}$, such that either $f(\hat{x}) = \lim\limits_{k\to\infty} f(x_k)$ or $f(\hat{x}) = \liminf\limits_{k\to\infty}(f(\hat{x}))$ and $f(x^*) - f(\hat{x}) \le G^2 h / 2$.

Proof: If $g(x_{k^*}) = 0$ for some k^*, then $f(x_k) = f(x^*), \forall k \ge k^*$ and we may take $\hat{x} = x^*$. If $g(x_k) \ne 0, \forall k$, then $x_{k+1} = [x_k + hg(x_k)]^+$. Let $z_{k+1} = x_k + h_{k+1}g(x_k)$ (without projection). Observe that

$$\left\|x_{k+1} - x^*\right\|_2 \le \left\|z_{k+1} - x^*\right\|_2. \qquad (17)$$

This is true as when we project a point onto C, we move closer to every point in C. Now,

$$\left\|z_{k+1} - x^*\right\|_2^2 = \left\|x_k + hg(x_k) - x^*\right\|_2^2 = \left\|x_k - x^*\right\|_2^2 + 2hg(x_k)^T(x_k - x^*) + h^2\left\|g(x_k)\right\|_2^2.$$

From (17), we have

$$\left\|x_{k+1} - x^*\right\|_2^2 \le \left\|x_k - x^*\right\|_2^2 + 2hg(x_k)^T(x_k - x^*) + h^2\left\|g(x_k)\right\|_2^2 \qquad (18)$$

From the definition of the subgradients for concave functions we have,

$$f(x^*) \le f(x_k) + g(x_k)^T(x^* - x_k). \qquad (19)$$

From (18) and (19), we get the following inequality $\square$

$$\left\|x_{k+1} - x^*\right\|_2^2 \le \left\|x_k - x^*\right\|_2^2 + 2h(f(x_k) - f(x^*)) + h^2\left\|g(x_k)\right\|_2^2. \qquad (20)$$

Recursively from (20), we get

$$\left\|x_{k+1} - x^*\right\|_2^2 \le \left\|x_0 - x^*\right\|_2^2 + 2h\sum_{i=0}^{k}(f(x_i) - f(x^*)) + h^2\sum_{i=0}^{k}\left\|g(x_i)\right\|_2^2. \qquad (21)$$

Using $\left\|x_{k+1} - x^*\right\|_2^2 \ge 0$, we have,

$$2h\sum_{i=0}^{k}(f(x^*)-f(x_i))\leq\left\|x_0-x^*\right\|_2^2+h^2\sum_{i=0}^{k}\left\|g(x_i)\right\|_2^2. \tag{22}$$

By property of concave functions, we have,

$$\frac{\sum_{i=0}^{k}f(x_i)}{k+1}\leq f(\overline{x}), \tag{23}$$

where, $\overline{x}=\dfrac{1}{k+1}\displaystyle\sum_{i=0}^{k}x_i.$ Thus we have,

$$\sum_{i=0}^{k}(f(x^*)-f(x_i))\geq(k+1)(f(x^*)-f(\overline{x})).$$

Combining this with (22), we get the inequality

$$2h(k+1)(f(x^*)-f(\overline{x}))\leq\left\|x_0-x^*\right\|_2^2+h^2\sum_{i=0}^{k}\left\|g(x_i)\right\|_2^2. \tag{24}$$

Given that $\|g(x_i)\|\leq G$, for all i, we have,

$$f(x^*)-f(\overline{x})\leq\frac{\left\|x_0-x^*\right\|_2^2+h^2(k+1)G^2}{2h(k+1)}. \tag{25}$$

Taking the limit as $k\to\infty$, we get,

$$f(x^*)-\liminf_{k\to\infty}(f(\overline{x}))\leq G^2h/2. \tag{26}$$

Hence the result.∎

6. REFERENCES

[1]   M. Gast, 802.11 Wireless Networks: The Definitive Guide. Sebastapol, CA: O'Reilly & Associates, 2002.

[2]   O'Hara and A. Petrick, IEEE 802.11 Handbook: A Designer's Companion. Standards Information Network, IEEE Press, 1999.

[3]   T. Nandagopal, T.-E. Kim, X. Gao, , and V. Bharghavan, "Achieving MAC Layer Fairness in Wireless Packet Networks," in ACM Mobicom, 2000, pp. 87–98.

[4]   S. Sharma, K. Gopalan, N. Zhu, P. De, G. Peng, and T. Chiueh, "Implementation Experiences of Bandwidth Guarantees on a Wireless LAN," in ACM/SPIE Multimedia Computing and Networking (MMCN 2002), 2002.

[5]   N. Vaidya, P. Bahl, and S. Gupta, "Distributed fair scheduling in a wireless LAN," in 6th Annual International Conference on Mobile Computing and Networking, 2000.

[6]   F. Kelly, A. Maulloo, and D. Tan, "Rate control in communication networks: shadow prices, proportional fairness and stability," in Journal of the Operational Research Society, vol. 49, 1998.

[7]   S.Kunniyur and R.Srikant,"End-to End Congestion Control Schemes: Utility Functions, Random Losses and ECN Marks," in INFOCOM (3), 2000, pp. 1323–1332.

[8] J. Mo and J. Walrand, "Fair end-to-end window-based congestion control," IEEE/ACM Transactions on Networking, vol. 8, no. 5, pp. 556–567, 2000.

[9] S.Srinivas and R.Srikant,"Throughput and fairness guarantees through maximal scheduling in wireless networks,"Network Optimization and Control, vol. 2, no. 3, pp. 271–379, 2007.

[10] X. Wang and K. Kar, "Cross-layer rate control for end-to-end proportional fairness in wireless networks with random access," in MobiHoc '05: Proceedings of the 6th ACM international symposium on Mobile ad hoc networking and computing. New York, NY, USA: ACM Press, 2005, pp. 157–168.

[11] J.-W. Lee, M. Chiang, and A. R. Calderbank, "Utility-optimal random-access control," Wireless Communications, IEEE Transactions on, vol. 6, no. 7, pp. 2741–2751, July 2007.

[12] J.Liu and A.Stolyar,"Distributed queue length based algorithms for optimal end-to-end throughput allocation and stability in multi-hop random access networks," in Proceedings of the 45th Annual Allerton Conference on Communication, Control, and Computing, September 2007.

[13] X. Lin and S. Rasool, "Constant-time distributed scheduling policies for ad hoc wireless networks," Decision and Control, 2006 45th IEEE Conference on, pp. 1258–1263, 13-15 Dec. 2006.

[14] A. Gupta, X. Lin, and R. Srikant, "Low-complexity distributed scheduling algorithms for wireless networks," INFOCOM 2007. 26th IEEE International Conference on Computer Communications. IEEE, pp. 1631–1639, May 2007.

[15] C.Joo and N.Shroff, "Performance of random access scheduling schemes in multi-hop wireless networks,"Signals,Systems and Computers, 2006. ACSSC '06. Fortieth Asilomar Conference on, pp. 1937–1941, Oct.-Nov. 2006.

[16] P. Chaporkar, K. Kar, X. Luo, and S. Sarkar, "Throughput and fairness guarantees through maximal scheduling in wireless networks," Information Theory, IEEE Transactions on, vol. 54, no. 2, pp. 572–594, Feb. 2008.

[17] C. Perkins and P. Bhagwat, "Highly dynamic destination-sequenced distance-vector routing (DSDV) for mobile computers," in ACM SIGCOMM'94 Conference on Communications Architectures, Protocols and Applications, 1994, pp. 234–244.

[18] J. Hsu and P. Burke, "Behavior of tandem buffers with geometric input and markovian output," IEEE Trans. on Communications, pp. 358–361, March 1979.

[19] R. Jackson, "Queueing systems with phase type service," Operations Research, vol. 5, pp. 109–120, 1954.

[20] K. Bharath-Kumar, "Discrete-time queueing systems and their networks," IEEE Trans. on Communications, vol. 28, no. 2, pp. 260–263, February 1980.

[21] D. Bertsekas, Nonlinear Programming, Second Edition ed. Athena Scientific, 1999.

[22] X. Wang and K. Kar, "Distributed algorithms for max-min fair rate allocation in aloha networks," in Proceedings 42nd Annual Allerton Conference on Communication, Control, and Computing, October 2003.

[23] N. Z. Shor, K. C. Kiwiel, and A. Ruszcaynski, Minimization methods for non-differentiable functions. New York, NY, USA: Springer-Verlag New York, Inc., 1985.

[24] N.Singh and R. Sreenivas, "Enforcing end-to-end proportional fairness with bounded buffer overflow probabilities," Coordinated Science Laboratory, University of Illinois at Urbana-Champaign, Urbana, IL, 1308 West Main Street, Urbana, IL 61801., Technical Report UILU-ENG-08-2211, August 2008.

Permissions

All chapters in this book were first published in IJWMN, by AIRCC Publishing Corporation; hereby published with permission under the Creative Commons Attribution License or equivalent. Every chapter published in this book has been scrutinized by our experts. Their significance has been extensively debated. The topics covered herein carry significant findings which will fuel the growth of the discipline. They may even be implemented as practical applications or may be referred to as a beginning point for another development.

The contributors of this book come from diverse backgrounds, making this book a truly international effort. This book will bring forth new frontiers with its revolutionizing research information and detailed analysis of the nascent developments around the world.

We would like to thank all the contributing authors for lending their expertise to make the book truly unique. They have played a crucial role in the development of this book. Without their invaluable contributions this book wouldn't have been possible. They have made vital efforts to compile up to date information on the varied aspects of this subject to make this book a valuable addition to the collection of many professionals and students.

This book was conceptualized with the vision of imparting up-to-date information and advanced data in this field. To ensure the same, a matchless editorial board was set up. Every individual on the board went through rigorous rounds of assessment to prove their worth. After which they invested a large part of their time researching and compiling the most relevant data for our readers.

The editorial board has been involved in producing this book since its inception. They have spent rigorous hours researching and exploring the diverse topics which have resulted in the successful publishing of this book. They have passed on their knowledge of decades through this book. To expedite this challenging task, the publisher supported the team at every step. A small team of assistant editors was also appointed to further simplify the editing procedure and attain best results for the readers.

Apart from the editorial board, the designing team has also invested a significant amount of their time in understanding the subject and creating the most relevant covers. They scrutinized every image to scout for the most suitable representation of the subject and create an appropriate cover for the book.

The publishing team has been an ardent support to the editorial, designing and production team. Their endless efforts to recruit the best for this project, has resulted in the accomplishment of this book. They are a veteran in the field of academics and their pool of knowledge is as vast as their experience in printing. Their expertise and guidance has proved useful at every step. Their uncompromising quality standards have made this book an exceptional effort. Their encouragement from time to time has been an inspiration for everyone.

The publisher and the editorial board hope that this book will prove to be a valuable piece of knowledge for researchers, students, practitioners and scholars across the globe.

List of Contributors

Ahmed Riadh Rebai
Electrical & Computer Engineering Program, Texas A&M University, Doha, Qatar

Saïd Hanafi
LAMIH Laboratory, University of Valenciennes & Hainaut-Cambrésis, Lille, France

Saurabh Goel
Software Engineer, Pariksha Labs Pvt. Ltd. India

NAJOUA ACHOURA
Department National Engineering School of Tunis, Tunisia

RIDHA BOUALLEGUE
SUP'COM, 6'Tel Laboratory,Tunisia

Debabrato Giri
Department of Information Technology, Jadavpur University, Kolkata, India

Uttam Kumar Roy
Department of Information Technology, Jadavpur University, Kolkata, India

S. Taruna
Banasthali University, Jaipur, Rajasthan

Rekha Kumawat
Banasthali University, Jaipur, Rajasthan

G.N.Purohit
Banasthali University, Jaipur, Rajasthan

Md. Mahedi Hassan and Poo Kuan Hoong
Faculty of Information Technology, Multimedia University, Cyberjaya, Malaysia

Salitha Priyanka Undugodage
School of Computing and Mathematical Sciences Auckland University of Technology, Auckland, New Zealand

Nurul I Sarkar
School of Computing and Mathematical Sciences Auckland University of Technology, Auckland, New Zealand

Jagbir Dhillon

Krishna Prasad

Rajesh Kumar

Ashok Gill

S Saravanan
Research Scholar, Dept of Computer Science, Bharathiar University, Coimbatore, India

E Karthikeyan
Asst. Professor of Computer Science, Government Arts College, Udumalpet, India

Dipak Wajgi
Department of Computer Science and Engineering, Shri. Ramdeobaba College of Engineering and Management, Nagpur, India

Dr. Nileshsingh V. Thakur
Department of Computer Science and Engineering, Shri. Ramdeobaba College of Engineering and Management, Nagpur, India

P. P. Bhattacharya
Department of ECE, Faculty of Engineering and Technology, Mody Institute of Technology & Science (Deemed University), Rajasthan, India

Ananya Sarkar
Department of ECE, College of Engineering and Management Kolaghat, West Bengal, India

IndranilSarkar
Department of ECE, Sobhasaria Group of Institutions, Rajasthan, India

Subhajit Chatterjee
Deparment of ECE, Swami Vivekananda Institute of Science and Technology, Barruipur, West Bengal

Dr. R. Shantha Selva Kumari
Department of Electronics and Communication Engineering, Mepco Schlenk Engineering College, Sivakasi

M. Aarti Meena
Department of Electronics and Communication Engineering, Mepco Schlenk Engineering College, Sivakasi

John Tengviel
Department of Computer Science, Sunyani Polytechnic, Sunyani, Ghana

K. Diawuo
Department of Computer Engineering, KNUST, Kumasi, Ghana

Jiang Liu
Global Information and Telecommunication Institute, Waseda University, 1-3-10, Nishi-Waseda, Shinjuku-Ku, Tokyo, 169-0051 Japan

Wasinee Noonpakdee
Global Information and Telecommunication Institute, Waseda University, 1-3-10, Nishi-Waseda, Shinjuku-Ku, Tokyo, 169-0051 Japan

Shigeru Shimamoto
Global Information and Telecommunication Institute, Waseda University, 1-3-10, Nishi-Waseda, Shinjuku-Ku, Tokyo, 169-0051 Japan

Nikhil Singh
Yahoo! Labs, Champaign, IL 61820, USA

Ramavarapu Sreenivas
Coordinated Science Laboratory & Industrial and Enterprise Systems Engineering, University of Illinois at Urbana-Champaign, Urbana, IL 61820

www.ingramcontent.com/pod-product-compliance
Lightning Source LLC
Chambersburg PA
CBHW080635200326
41458CB00013B/4640